This book goes beyond conventional wisdom about optimizing siting decisions. It illustrates the tradeoffs between economic, environmental and social objectives in a globalized world and provides best practice advice on how to deal prudently and responsibly with uncertainty and risk in locating hazardous facilities. A must for all planners, managers and regulators!

Ortwin Renn, Scientific Director, Institute for Advanced Sustainability Studies, Potsdam, Germany

Siting Noxious Facilities

Siting Noxious Facilities explains and illustrates processes and criteria used to site noxious manufacturing and waste management facilities. It proposes a framework that integrates economic location analysis and risk analysis, emphasizing the reduction of uncertainty.

This book begins by defining noxious facilities and considers the important role of manufacturing in the world economy, before going on to describe the historical practices used in locating these facilities for much of the twentieth century. It then shifts focus to analyze the complex set of considerations in the twenty-first century that mean that any facility that produces annoying smells and sounds, is unsightly and emits hazardous substances has had the bar of acceptability markedly raised for economic, environmental, social and political acceptability.

Drawing on case study examples that highlight pollution prevention, choosing locations at major plants (CLAMP), negotiations, and surrendering control of an activity, Greenberg presents a hybrid framework that advocates the amalgamation of industrial location processes with human health and environmental-oriented risk analysis.

This book will be of great interest to students and scholars of location economics, environmental science, risk analysis and land-use planning. It will also be of great relevance to decision-makers and their major advisers who must make choices about siting noxious facilities.

Michael R. Greenberg is distinguished professor and interim dean of the Edward J. Bloustein School of Planning and Public Policy, Rutgers University, USA.

Earthscan Risk in Society
Edited by Ragnar E. Löfstedt
King's College London, UK

Siting Noxious Facilities

Integrating Location Economics and Risk Analysis to Protect Environmental Health and Investments

Michael R. Greenberg

Routledge
Taylor & Francis Group
LONDON AND NEW YORK

earthscan
from Routledge

First published 2018 by Routledge

2 Park Square, Milton Park, Abingdon, Oxfordshire OX14 4RN

52 Vanderbilt Avenue, New York, NY 10017

Routledge is an imprint of the Taylor & Francis Group, an informa business

First issued in paperback 2020

British Library Cataloguing-in-Publication Data
A catalogue record for this book is available from the British Library

Library of Congress Cataloging-in-Publication Data
Names: Greenberg, Michael R., author.
Title: Siting Noxious Facilities: Integrating location economics and risk analysis to protect environmental health and investments / by Michael R. Greenberg.
Description: Abingdon, Oxon; New York, NY: Routledge, 2018. | Series: Earthscan risk in society series | Includes bibliographical references and index.Identifiers: LCCN 2018003837 |
ISBN 9781138099654 (hbk: alk. paper) | ISBN 9781315104034 (ebk: alk. paper)
Subjects: LCSH: Industrial ecology. | Factories–Location–Economic aspects. | Industrial sites–Environmental aspects. | Social responsibility of business. | Environmental protection. | Factory and trade waste–Environmental aspects.
Classification: LCC TS161.G7535 2018 | DDC 628.4–dc23
LC record available at https://lccn.loc.gov/2018003837

ISBN: 978-1-138-09965-4 (hbk)
ISBN: 978-0-367-50767-1 (pbk)

Typeset in Goudy
by Wearset Ltd, Boldon, Tyne and Wear

To our wonderful grandchildren: Lola, Layla and Luna Suggs; Amelia, Max and Zane Wilkerson; Faith Pappas; Willie and Eli Jacobs-Perez. You make me smile and laugh.

Contents

PART III
Tools and coping with siting noxious facilities in the early
twenty-first century   **213**

Illustrations

Figures

Tables

Boxes

Preface

When I was a young boy, I had seasonal pollen allergies and asthma. I hated those allergies. I sneezed, coughed; and they limited what I could do during the spring and fall months. I could neither smell nor taste normally. However, when I turned 15, not being able to smell became a positive attribute. I took a summer job working for a printer located in lower Manhattan. The printing plant had lead type, paper cutters, printing machines to name the three most obvious hazards to a teenager. The plant odors bothered my co-workers, but being 15 and unable to smell normally, I was clueless. I collected my hourly pay of $1 an hour and felt a sense of independence.

What my olfactory deficit could not protect me from was the factory in the next building that manufactured leather clothing. Every time we went out for lunch or a break, the noxious fumes hit us, even me, with an acrid nauseating smell. It was then that my co-workers told me that I was lucky to have those allergies – the smell was worse for them than it was for me. It was their full time job, so they put up with it. But even though it was only a summer job for me, the leather clothes factory really bothered me. I couldn't understand why this stinky factory would be allowed to fill this lower Manhattan area with noxious fumes.

I tried without success to figure out how that plant got there and how many others were in New York City. I also had the opportunity to learn about slaughter houses, candle and glue factories, and a steel plant that smelled, and I learned that some of them discharged wastes into the sewers or nearby streams. Zoning didn't necessarily keep these facilities out of light manufacturing, commercial or even residential districts. In my undergraduate chemistry classes, I learned that some discharges from these noxious factories were bad for humans and the environment. With this information, I asked my undergraduate college economics professor about how New York City could allow the land values of lower Manhattan to be reduced by this noxious plant; his answer demonstrated to me that he knew a lot more about macroeconomics than microeconomics.

My ability to understand how and where noxious facilities were located markedly increased in 1965 in graduate school when I took a course in location theory taught by Professor Herman Otte who had a joint appointment in the business school and geography department at Columbia University. Professor

Otte had studied with German location theorists and could explain and illustrate all of the major location theories. Much of the literature was written in German, little had been translated, and so I was exceedingly fortunate to have this outstanding professor and to be able to read German (with a dictionary at my side). Dr. Otte believed in what he called "location science," and he convinced me that it was a real science (Chapter 2). In fact, after he retired, I was hired by Columbia University and taught a course in location theory. I was convinced that profit maximization and cost minimization drove the location process. By the time I understood those theories, they were out of date.

By 1970, I no longer believed that the location science I had learned and was teaching worked in the real world. Money still counted more than anything else, but President Richard Nixon, a Republican with a conservative platform, signed the National Environmental Policy Act of 1970 (NEPA), European nations followed, in fact, over 100 nations had their own versions of a national environmental policy law. In the United States, NEPA was followed by the Clean Air Act, the Clean Water Act, and a host of others (Chapter 4). I did not see how many of the noxious production and waste management sites that I knew about were going to survive this legislation and public opposition to noxious facilities. I was certain that post-1970, location analysts would need to rethink where to site manufacturing and waste management facilities that produce annoying odors, visible emissions, noise, and expose workers and people to toxins.

My education about noxious facility siting increased when I moved to New Jersey in 1971. I watched old industrial location theory literally disintegrate before my eyes. I lived 10–20 miles from what had been more than a dozen economically viable chemical, automobile assembly, smelting and other industrial facilities that epitomized traditional economic-grounded industrial location science. I watched these rock solid manufacturing industries decline, close and be accused of leaving intolerable human health, safety, and environmental legacies (see Chapter 3). The dramatic on-the-ground changes were hard to grasp. When I was a graduate student, Leonard Zobler, one of my two major advisers, said that "It may be brown and gray to you, but it is green [meaning profits, jobs, and tax revenues] to them [meaning owners, workers, and city officials]." By 1980, Leonard's expression was not acceptable to many people. Watching protests, law suits in my own state and many others, I recognized that we had entered the realm of psychology and sociology as critical industrial location theory factors.

Human health and the environmental challenges were a reason that these plants failed, but not the only ones. The second major reason for my loss of confidence in the location theory I had learned and taught was the changing world economy (Chapter 1). We had been taught that the best location was almost always going to be near key resources, customers, workers, and in a dense transportation network. When Japan and China began to sell steel, cars, and copper in New York and Chicago, I knew this was a new world economy. Enter globalization and the economic, social, political and environmental challenges it brings.

I consider myself a realist looking for practical workable options while avoiding the trap of immobile ideological-grounded positions. Hence, despite strong economic, social, health, environmental, moral and political labels attached to noxious facilities, my premise is that they are not going to be an endangered land use species in the foreseeable future. Private and public mangers are going to need to find a pool of tolerable sites for these sites, places where they can be expanded or and provide non-sting options.

Like others who write about policy topics, I am a captive of time and space. The time span for this book is the century that begins with the end of World War I, when the United States was just about to become an urban nation. I reflect and interpret what I have seen in North America and to a lesser extent in the Ruhr Valley and British Midlands of Europe. I have visited countries in Asia, but have not had any more than limited conversations with people involved in industrial locations, many of whom went out of their way not to show me noxious facilities or talk about how they were located. In short, my personal experiences are grounded in only two of the world's most important industrial clusters.

While acknowledging my limitations, I know that industry and waste managers have not renounced economic efficiency and the profit motive. What has changed is that they face new layers of complexity, and what I see them doing is trying to increase their control and reduce uncertainty imposed by exogenous factors that they can minimally influence, if at all (Chapter 4).

In the early twenty-first century, managers and their organizations walk on a tightrope trying to balance siting and non-siting solutions. Explicitly connecting economic location and health and environmental oriented risk analysis can lead to improvements in protecting both environmental health and economic investments. Managers must confront advantages and disadvantages of siting and non-siting options, weigh uncertainties with each option, and think about these as best they can over the life cycle of the proposed facility (Chapters 5–10). And their use is going to require thinking about what is a realistic set of noxious facility siting guidelines for today and tomorrow, not for two generations ago, as well as incorporating processes and analytical tools into economic and risk analysis processes that allow different kinds of data to be collected and included in decision-making.

The broad goal of this book is to tell the roughly century-old story of change in the process of locating noxious facilities from traditional economic-grounded location principles toward managing uncertainty by using both economic location and risk analysis frameworks and tools. I make no claim that the subject matter, arguments and tools presented in this book are new, although I have sharpened some of them. What is different is that they are in one volume, not in three, and I connect them. Specifically, as discussed in the book, there is a classical economic-based location literature that has been tweaked during the last two decades (Chapter 2). Second, there is a large post-1970 literature about siting industrial and waste management facilities with a focus on environment, equity and autonomy issues (Chapter 4). Then there are other reports, again typically disconnected from the first two, about how to make noxious facilities less noxious using various forms of pollution prevention (Chapter 5).

After thinking about this book for over a decade and speaking with professionals who work in this field, academic colleagues and current and former students, I wrote this book for three audiences. One is the set of decision-makers who need to be able to balance all the layered factors that impact on siting decisions. By far they have the most difficult job. They need to know how the key factors driving siting options intersect, how uncertainty can be reduced, and how they can make choices that make noxious facilities less obnoxious. Second, I have worked with experts who know a great deal about economic factors, others who understand the human health and environmental challenges, and those who are primarily motivated by social and political outcomes. This book, I hope, will provide them with context for their work. Third, current students will face an even more complex web of factors that must be balanced. For them, the book tries to drive home the reality that the encompassing location science as I learned it is dead. It has been replaced by siting processes that are as much art as science. The challenge is balancing many factors and trying to reduce uncertainty to protect major investments while at the same time enhancing public health and the environment, or at least not worsening them. Those intending to work in this noxious facility siting craft must not be faint of heart. They must be able to go head to head with managers, stockholders, elected officials, environmental organizations, and residents that have core values tied up in siting decisions and will express these often not in the most flattering of terms, if you happen to disagree with them.

In regard to the design of the book, over a half century of teaching and writing, I have learned that readers want real examples to guide them, and do not want to choke over equations. Hence, I have provided illustrations throughout the book, and with a few exceptions I have deliberately deemphasized quantitative analytical tools, although I have provided references to these. Also, as much as possible I have chosen examples that I am personally familiar with, that is, worked on and visited on multiple occasions. When I did not have personal history, I picked examples that I have access to through colleagues and a black and/or gray literature. Of course, that limits the range of examples, but writing about a case based solely on popular portrayals or even the black literature is not a comfortable option for me. Being there, seeing it, and discussing the site are critical. Hence, with a few exceptions the major examples are from the United States and the non-U.S. illustrations receive much briefer treatment.

If I have done a good job writing this book, the examples will demonstrate what factors drove siting in the past compared to the evolving present. I hope that the book will stimulate your interest in what has become a fascinating and difficult challenge, one requiring a high level of multidisciplinary knowledge and the time and patience to apply it in front of communities that may not be receptive. The cover of my 2016 book *Explaining Risk Analysis* features a man walking on a tightrope over a body of water with sharks swimming around, which is the image I have for those of you who are responsible for making the choices.

Michael Greenberg

Acknowledgments

As a graduate student at Columbia University I was fortunate to have advisors who cared about location theory and about me. I thank my major advisors George Carey, Douglas McManus, Herman Otte, William Vickrey, and Leonard Zobler for spending many hours coaching me about location theory and becoming a professor. Since moving to New Jersey, I have had mentors that have taught me a great deal about risk and allowed me to work on some of the most interesting location projects in the world, some of which are described in the case study chapters and others of which are still in play. I appreciate the opportunity to work with Jim Hughes, Joe Seneca, Bernard Goldstein, Henry Mayer, Paul Lioy, and Arthur Upton of Rutgers University, David Kosson and Charles Powers of Vanderbilt University, and Mark Gilbertson of the U.S. Department of Energy.

In regard to preparation of the book, Diren Kocakusak helped with the international examples in Chapters 5–8. Tamara Swedberg drew or redrew nearly all the maps and charts. I used photos to help explain the geographical attributes of some sites, especially in Chapter 3. I thank ESRI for having these photos available. I thank my wife, Gwen Greenberg, for drawing the picture of the Manville site as I remember it.

The opinions, findings, conclusions and recommendations expressed in this book are mine and do not necessarily represent the views of anyone listed.

1 Introduction to siting noxious facilities

Introduction

This chapter has three goals:

- Describe the changing geography of world economic growth, especially the clustering of hot and cold economic areas, and review how these changes impact the location of noxious industries;
- Provide a working definition for noxious facilities for this book; and
- Explain the organization of the book.

The backdrop: world economic growth, clusters of hot and cold economies, and manufacturing

If you type Google Time Lapse into your browser, and then type in Shanghai, Hanoi, Ankara, or other rapidly growing cities, you will witness an amazing visual display of clustered growth that took place between 1984 and 2015. Agricultural lands turn into developed settlements, new roads and bridges appear, as do airports and other urban activities. Factories appear in clusters near transportation sites, and then you see disturbed land, which sometimes means that you have found a landfill, a quarry or a site under construction. What you see in these images is the backdrop for finding new production sites, clamping on to existing sites, and developing non-siting options for noxious facilities that emit contaminants into the air and water, are unsightly, and have the potential to cause health impacts.

A growing and globalizing world economy

The United States was the biggest actor on the economic stage in 1965 when I took my first location theory course. The U.S. economy accounted for about a third of the world's economic output. But the world economy grew about eight times (author estimates from published data in constant dollars) between 1965 and 2015.[1,2] Slowly and then more rapidly, Japan and Europe rebounded from World War II, and then China, India, Brazil and other states began to grow economically. Now we know that no one country, not even the United States, can set the world's economic agenda using its own rules and processes. The world is heading toward a more global economy, despite some political efforts to slow it down.

More transportation options are one reason. Production facilities need not always be located near raw materials and/or in markets. In 1965, cargo ships were able to move crude oil, some manufactured products, and people across the globe. In 2018, long distance transport capacity is markedly enhanced and much more is being moved and arriving at the customer's door at cheaper prices. For example, in 1965, I could not have imagined that it would be possible to compress natural gas to 1/600th of its volume and then ship it in tankers the size of aircraft carriers. Cargo ships are getting bigger, posing challenges for ports and requiring dredging to depths not considered in the past. Yet, the need is being accommodated in order to move raw materials, semi-finished and final manufactured products. Manufacturing continues to be at the heart of stimulating economic growth, as scholars asserted over a half century ago (see Chapter 2).

Enhanced communications are a central part of globalization, meaning that entrepreneurs, scientists, and engineers can explore opportunities with partners all over the world at any time. I am sure that everyone reading this book has had to deal with someone trying to hack into their Internet accounts. Securing communications has become a major global industry and a new consideration for decision-makers. While we worry about trade secrets and military technologies,[3] we should also worry about our capacity to protect the operating systems of noxious facilities from cyber attacks.

The world's financial system is increasingly global. A financial investment in one country will be reflected, sometimes very quickly, in stock prices and political decisions elsewhere. A failed supply chain in one country can lead to an inability to produce what is needed in another, which can be a fiscal upset for workers and towns that depend on those facilities for jobs and taxes. When these upsets occur, businesses, local governments and populations that depend on internationally owned production facilities are in a precarious position with little recourse.

Globalization has undermined the traditional North American and European manufacturing job, the kind held by the media's creation Chester Riley in a California aircraft plant (see Chapter 2). High wage and high benefit unionized manufacturing jobs have been jeopardized by robots and a global labor market. In regard to noxious industry, manufacturing jobs may not seem as desirable as they were, especially if there are health and safety risks associated with these jobs. Globalization, however, has benefited extremely skilled workers who can move into high paying jobs in the most desirable locations, at least until someone or some machines can replace them.

In regard to human health, safety and environmental protection, the world population increased from 3.3 billion in 1965 to 7.2 billion in 2015. Many people live in relatively safe and clean environments and can expect to live well into their 80s, with an increasing number living more than a century. But in other places, life expectancy and quality of life have deteriorated. Inequities by income, nationality, race, ethnicity and age are striking both globally and within countries.[4]

In 2014, world gross product (WGP), which combines the gross national product of all nations, was closing in on $80 trillion (U.S.$)[5,6] This world gross product is an increase of more than 30 times in a century.[1] Barring major financial, political, military and/or natural hazard events, the world economy will continue to grow. The Conference Board[2] expects growth rates of 3.0% annually between 2017 and 2021 and 2.7% from 2022 through 2026. These estimates imply enormous ongoing needs for manufacturing and waste management sites.

Noxious industry and waste management are only part of the large, complicated, and rapidly growing world economy. But because of their historical association with worker exposures, distasteful odors, loud sounds, unattractive visuals, as well as toxic emissions, these industries are the business sector clearly under the public and private magnifying glass for health and environmental risk assessment and management. Globally, I believe that we are in for a long period during which every decision regarding noxious facilities is going to be heavily scrutinized. To be the owner and/or manager of a noxious requires continuous readiness for unfriendly scrutiny.

Clustering of hot and cold economies

A view of the world economy as a whole does not capture the formation of hot and cold economic clusters (Figure 1.1). Some countries, most obviously, China have been growing at an accelerated rate, even though the rate of growth has slowed down (Table 1.1). In addition, Indonesia, Nigeria, Saudi Arabia, Thailand, Venezuela, Vietnam, and several others have been growing at more than 5% a year.[2,5,6] Other countries have not been growing, at least in the recent past.

Figure 1.1 Key world economies.

Table 1.1 Size of the world economy and manufacturing sector

Country	Gross domestic product, trillions 2014	Gross domestic product, 2014, % world	Manufacturing output, 2014, % world (rank 1–10)	% manufacturing of national economy, 2000	% manufacturing of national economy, 2014
United States	17.3	22.2	17.8 (2)	16	12
China	10.3	13.2	26.4 (1)	32	30
Japan	4.6	5.9	7.5 (4)	21	19
Germany	3.9	5.0	7.7 (3)	23	23
United Kingdom	3.0	3.8	2.8 (5)	16	11
France	2.8	3.6	2.6 (8)	16	11
Brazil	2.4	3.1	2.5 (9)	15	12
Italy	2.1	2.7	2.7 (7)	20	15
India	2.0	2.6	2.7 (6)	15	16
Russian Federation	2.0	2.6	1.0 (10)	17	6
Remainder of world*	27.7	35.4	26.4	NA	NA

Sources: Conference Board,[2] World Bank,[5] Central Intelligence Agency.[6]

Note
* Includes 204 places, many of which present no data because the data are subsumed in another national data set.

Italy, Spain, the Netherlands, Belgium and Denmark experienced negative growth rates in 2012, for example.

More broadly, there are four major world economic growth clusters. The first includes a set of European Union nations, with Germany, the United Kingdom, Italy, and France as the largest. Second, the United States, Canada, and Mexico form a North American cluster. The Asian cluster, containing much of the world population, is more geographically dispersed, including China, India, Japan, South Korea and Indonesia. Brazil leads a smaller and developing fourth cluster of South American nations that also includes Argentina, Venezuela, and Columbia. These four clusters accounted for nearly all of the world growth in gross domestic product between 2000 and 2014, especially in manufacturing (see Table 1.1).

Forecasters expect China to replace the United States as the number one economy measured by GDP in the foreseeable future, probably around the mid-2020s.[2,5,6] The CIA, World Bank and other estimates for 2020 through 2025 show annual GDP growth rates of 3.5%+ in China, India, and other developing parts of Asia. Mexico, and parts of Sub-Saharan African are also predicted to have a growth rate of 3% or more. By comparison, the rates of United States, the European Union, and Japan for this same period are estimated as 1.7%, 1.2%, and 0.6%, respectively. Assuming these changes occur, the focus of industrial and waste management locational analysis, will follow.

The world economy has relatively cold economies, exemplified by large parts of Africa, Asia, and South America that have growing populations but low per capita income generation. The most obvious are in Africa, including the Democratic Republic of the Congo, Niger, Zimbabwe, Somalia, Burundi, Liberia, Eritrea, and the Central African Republic. Beset by the AIDs epidemic, wars, a lack of food, sanitation and water problems, medical and other services, they have incomes less than $1000 per year.[2,5,6]

As a visitor to some of the relatively poor and slow or declining economies, I know that there is a need, desperate in some cases, for new schools, hospitals, housing, power, water and sewer connections and plants, and many other structures. When a major worldwide negative economic event occurs, such as the recession that began in 2007, the media emphasize the impacts on the northern hemisphere urban-industrial nations. However, these events markedly reduce the chances that poorer nations, typically in the southern hemisphere, will receive funds for these new facilities.

Globally, the economic geography of gross regional product tells us where noxious industry will concentrate. In addition to standard economic reporting[2,5,6] special reports focus on specific sectors. One of the most important is the forecast of the growth of the world waste management industry,[7] which is expected to experience substantial growth, especially in the Asia-Pacific area because of population and industrial growth, as well as the reality that with income growth these rapidly developing countries will be able to afford to upgrade their inadequate waste management systems.

Within each nation, some areas are growing rapidly and others less so, if at all. Growth in the United States, for example, has been trending toward the

South and West for many years, but more recently growth is shifting back from suburbs to cities, and now to areas that are rich in energy resources. In China, massive state expenditures have led to metropolitan clusters in Beijing, Shanghai, and to the south and west of the country, such as Wuhan, Changsha, and Chongqing where the government has chosen to enhance urban clusters. In India, cities like Mumbai, Bangalore, Hyderabad, and Ahmedabad have experienced massive growth.

Because of their sheer size, urban agglomerations in China and India attract a great deal of attention, and they should. Yet there is also Lagos, Mexico City, Karachi, Manila, Seoul, São Paulo, Moscow, Jakarta and other cities of over 10 million where public and private managers who direct location analysis work on these are under tremendous pressure to make good choices. I believe that decisions about noxious facilities will prove to be among the most difficult as the populations in many countries become less comfortable with noxious facilities (see below and Chapter 10 for a presentation about garbage disposal).

The United States and China: contrasts and similarities

A great deal can be learned about industrial location trends, including about noxious industry, by studying some of the most rapidly industrializing nations, which would include Nigeria, The Democratic Republic of the Congo, the Czech Republic, Vietnam, and semi-independent places like Puerto Rico before the fall 2017 hurricanes. However, I have not visited all of these countries or have spent only a limited amount of time in them. I am reluctant to focus on places that I have not spent much time in. Accordingly, here I compare the United States and China in regard to manufacturing trends.

China and the United States are the top dogs in manufacturing and yet at a different stage in their development, including their manufacturing sectors. These two combined for 48% of world manufacturing production in 2000 and 42% in 2014 (Table 1.1). There is a common belief that the United States is hemorrhaging manufacturing and that China continues to add it. A quick glance at Table 1.1 shows that the United States continues to have a lot of manufacturing jobs and the role of manufacturing in China is being reassessed by its leaders. Because siting decisions follow industrial policy decisions, I briefly review the role of manufacturing in these two massive economies. A caveat is that public information about the behind the scenes reasons for the patterns that exist are limited. Indeed some involve national security concerns and trade secrets.

The United States

There is no doubt that the U.S. has lost some of its manufacturing base. However, it is not true as I heard during the 2016 presidential election in the U.S. that "we don't manufacture anything anymore." The Global Macro Monitor,[8] which is a source for investors, traders and policy makers, published a one-page "chart of the day" about U.S. manufacturing employment. Half of presentation was a chart of U.S. manufacturing jobs from 1960 to March 2012. The chart shows that there was an upward trend from 14.8 million jobs in 1951 to 19.5 million in 1977, when manufacturing accounted for 22% of the nonfarm payroll. Beginning in 1998, manufacturing jobs began to decline, reaching 16 million in 2012, precisely what they were in the year 1950. So while it is true that the United States has not abandoned manufacturing, it certainly has not been growing manufacturing for four decades. The authors of the report attribute the decline to the following factors:

- Strengthening of the value of the U.S. dollar during the 1980s, which made U.S. products relatively more expensive compared to competitors;
- Globalization;
- Growth of China and India as competitors;
- The Internet;
- Productivity increases;
- Technology innovations;
- Demographics and worker preferences; and
- All of the above.

This somewhat incomplete list ends with the following provocative statement: "We'll leave it to the academics to debate it and the politicians to place blame or take credit."[8] In this book, I will not debate the causes of it. I would rather challenge the implicit assumption that losing manufacturing is bad and growing it is good. For me, the real issue is what is in the best interests of the U.S. as a whole and its constituent regions and populations, which I believe is not a simple question to answer. For example, one of the debates in the United States is should the U.S. export natural gas via LNG tankers to Japan, parts of Europe and elsewhere. The United States already exports natural gas in North America as required by its treaties with Canada and Mexico. Some national scale, economic simulations show that the entire economy benefits as a whole from more gas exports. But this assertion has been challenged by other simulations that argue that only the oil and gas industry would benefit and that Midwestern manufacturing industry would be hurt by higher gas prices.[9–13] The real issue is who gains and who loses, individually and regionally.

Another important question is how much do Americans want to pay for clothes, refrigerators, and other durable products? If it means that we will need to pay much more for these by manufacturing them in the United States, does it make sense to try to recapture those manufacturing specific industries? The author's family physician for more than 40 years set up a small foundation to investigate the benefits of bringing back industry to the United States, assuming that Americans would be willing to pay more for every shirt than they have been. He was discouraged to find from survey data that only a small proportion of people he studied were willing to have a smaller wardrobe. A clear plurality wanted more clothes and lower prices.

What would the United States need to do to regain more of industries that make sense for specific regions and populations? What tradeoffs would be required? With the British vote to exit the EU, the United States vote for Donald Trump who stated that the U.S. needs to stop off-shoring manufacturing jobs, and similar political issues stated in other European nations, there is renewed discussion about manufacturing. But these exchanges are much more about political positioning and become a tiresome distraction rather than a helpful step in making informed and realistic policies.

What do these political debates mean for manufacturing in general and noxious industries, more specifically, is the concern of this book. For me, the issue of recapturing industry is a mix and match one. What is the right mix of industry and what places are likely to support those manufacturing processes? United States' trade and currency policies have not supported manufacturing retention and expansion. Writing for the Economic Policy Institute, Scott[14] argues for rebuilding manufacturing. Adopting the principle of economic base theory (see Chapter 2), but not calling it that, he argues that manufacturing is the most important sector. He notes that the U.S. manufacturing sector employed 12 million people in 2013 (8.8%), and that another 17.1 million indirect jobs, in other words, over 29 million are associated with manufacturing in the United States, or 21% of the national employment. He adds that manufacturing jobs pay higher wages and have better health benefits than do most service jobs, and that this is particularly important for people without a college education.

Should the United States try to regain jobs lost to China, Mexico, South Korea and other countries during the 2007–2008 recession and earlier? Your answer likely depends upon where you sit. If you live or are heavily vested in Indiana, Michigan, Ohio, Wisconsin, Iowa, North Carolina, South Carolina, Alabama, Kentucky, Mississippi, Tennessee, Arkansas, and Oregon you are likely to want the United States to prioritize manufacturing in international negotiations.

Scott argues that the United States has allowed China, Japan and selected other counties to manipulate their currency. This manipulation, he says, has placed U.S. industry at a competitive disadvantage and this has not only raised the trade deficit but also undermined places that depend on manufacturing.

As I watched presentations made by the United States' presidential candidates in 2016, it seemed as if I was watching shows about China-bashing that sometimes were tied to manufacturing. Several of the presidential candidates seemed far more knowledgeable than others about China's role in the world economy and the impact on the U.S. economy, but all seemed intent upon blaming China rather than focusing on the United States. Commenting on the discussion of manufacturing in the U.S. presidential elections, Rothfeder[15] pointed to data that manufacturing has actually been increasing in the United States since 2010, mostly in the South and that manufacturing jobs in China have been decreasing since 2012, a point that was not a secret during the presidential debates, but did not seem to be mentioned.

So-called "reshoring" of some jobs that primarily were in China and Mexico has been occurring. Contributing factors include increasing wages in China, new oil and gas supplies in the United States, a high quality U.S. workforce, more reliance on technology (hi-tech equipment), and flexible scheduling for workers. A fascinating part of this reshoring is that manufacturers want to be as near to their clients as possible so that products can be quickly adjusted to the market and delivered to rapidly changing markets. In the 1960s, we called this market-oriented manufacturing, focusing on transportation costs. Today, it is to have customers nearby to provide input to designs. Reshoring will, however, not necessarily help those who were in industries that are in labor cost sensitive industries such as a typical T-shirt manufacturer.

Some U.S. business leaders have been pressing for more effort to grow manufacturing. Jeffrey Immelt, then the CEO of General Electric, a company that has been severely criticized for off-shoring and polluting the Hudson River, has argued that manufacturing that requires innovation, excellent workers, agile responses to customer needs, and is near resources should return to the United States. Lower labor costs are not a priority for these types of manufacturing because there is not so much labor time in the production process. In GE's case this means manufacturing refrigerators, jet engines, turbines, and MRI scanners. Immelt has heavily invested in research and development, which many U.S. businesses have not historically done. Immelt[16] does not forecast a return to 20% of U.S. jobs, but he sees steady growth up from 9%.

Denning[17] argued that Immelt is about making money, wherever he can and not about being a patriot. Investing in the U.S., Denning argues will happen when it benefits the companies and their stockholders, and meanwhile businesses like GE will continue to destroy U.S. jobs and give away intellectual property to increase the profits of their shareholders.

I do not take what Immelt or other business leader say at face value.[18,19] What I do see in Immelt's remarks and behaviors is a path forward about what one durable manufacturer who manages noxious industries in his portfolio believes to be a way of adding manufacturing jobs in the United States. In his 2012 article in the *Harvard Business Review*, Immelt[16] noted that the outsourcing is not over, but that a new era can return high paying jobs to the U.S. based on high technology, a lean and well educated workface, investment in education and research, locations where customers are nearby, and a labor force willing to understand what compensation packages can be in these jobs. He also adds that the U.S. needs to upgrade its infrastructure, support education, and he provides examples of where his company has done both.

> What GE is doing … across the United States says something about how we make decisions concerning where to manufacture our products.
>
> (Immelt,[16] p. 2)

Immelt seems clear about his expectations. What this implies for noxious industries in unclear because the discussions are at a high level of abstraction (see Chapter 5 for an interesting illustration of Merck).

China

China is trying to accomplish in one generation what Western Europeans and North Americans had three to eight generations to accomplish, that is, building health, education, water, sewer, and many other systems and in general raising the quality of life of their populations. China was an extremely poor country, invaded and devastated on multiple occasions, especially during World War II. With its leaders responsible for a population of more than a billion people, China has chosen to do whatever it needed to do to grow, using manufacturing as was the case in the United States and parts of Europe as their basic industry that would allow the government to turn manufactured products into jobs and revenues.

Some people and groups are outraged and charge that China does not respect intellectual property, has manipulated its currency, and so on. Whatever your views of the morality of its practices, China has not been the only country to be accused of unethical practices, including the United States. Whatever their ethical practices, China has established a massive manufacturing base in the northeastern China coast (Jiangsu, Guangdon, Shandong, Shanghai and Zhejiang provinces). The model establishes manufacturing, builds infrastructure, uses these to build more manufacturing, attracts innovators to build more manufacturing, infrastructure and trade and attracts external resources to improve quality of life. It bears a striking resemblance to what was done in Europe and the United States, as summarized in Figure 1.2 (see Chapter 2 for more detail).

China surpassed the United States in manufacturing in 2011 and became the world's largest exporter of manufactured goods. For more than three decades, China's economy grew at more than 10% a year, nearly all of it driven by its commitment to manufacturing. The net benefit has been to raise the quality of life, but not equally across the country. It also has been investing in rail and other infrastructure to better link and raise the standard of living in southern and western parts of the country.

Visitors to the country, including this author, have clearly seen some of the problems associated with this expansion. Obvious ones are massive air pollution in Beijing and several other cities with more than 10 million people, and unfinished infrastructure across the nation, including in the middle of large cities. While Westerners may not appreciate some of the pollution they see in China, I suggest that Chapter 3 shows similar or even worse problems in the United States beginning at the turn of the twentieth century. China vigorously embraced the idea of export base theory focused on industrial development (see Chapter 2), and created massive industrial clusters that are larger than those in New York City, Chicago, Philadelphia, Pittsburgh and Cleveland a half century earlier. What is hard to grasp, even during visits, is the magnitude of the social, health, environmental and political challenges facing Chinese decision-makers.

Social strife is another issue that arises in places in transition between an agrarian and industrial economy. The author vividly recalls the building of major highways through the middle of established urban neighborhoods in the

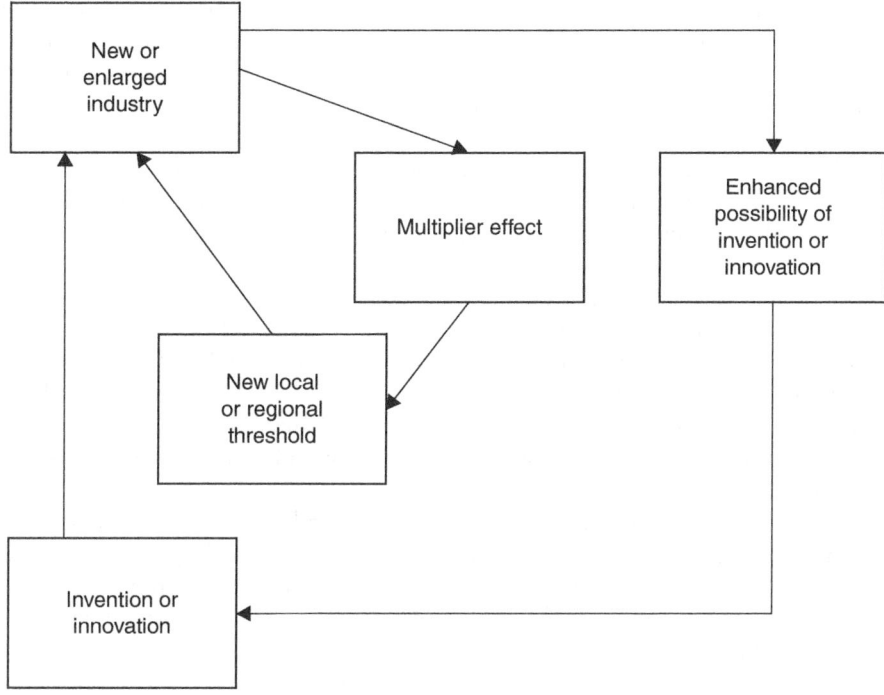

Figure 1.2 Simple model of development through manufacturing.

Bronx, New York. Dynamite blasts and relocation were the norm for some years close to our apartment in the south Bronx when for better or worse New York City built the Cross Bronx Expressway (I-95).[20] Now planners and engineers are talking a about building parks on top of the large highways to improve quality of life along the roadway.

The Chinese government has to deal with many such projects and of larger magnitude. Earlier, New York City built a massive aqueduct that brings water over 100 miles. China is building a much larger version to supply the growing areas of north China. My personal experience is that the Chinese government is acutely aware of these problems, has reached out for advice that it believes not to be politically motivated and has taken action to address what it can deal with given this historical legacy of poverty and inequity across its landscape.

I am neither an apologist nor proponent for all that I have seen and read about in China. The questions I have been asked by Chinese government environmental ministry officials and academy of sciences officials appeared genuine and all of these led to give and take. I spent considerable time responding to questions about energy sources, other noxious industries, use of risk analysis, and the role of manufacturing in stimulating the economy. Most frequently, I was asked how could China reduce coal use and how risk analysis could be used to help decision-making on this and other issues, such as phasing out old trucks and cars, locating facilities where domestic and hazardous waste could be properly managed. The Chinese government has embraced the concept of social risk assessment for large industrial projects.[21] How this translates to action on the ground is not clear other than it has been portrayed as an effort to reduce protests.

Industry and construction have accounted for almost half of China's gross domestic product. It became the largest steel producer in the world, third in automobile production, and in a variety of electronic and machinery final products and components, as well as cement and fertilizers. Indeed, looking at Chinese industrial production data, I am hard pressed to identify a major sector that it is not among the leaders from small toys to large rail cars. With a manufacturing growth rate of more than 10% a year, China served as the place where offshore industrial location, including much of the U.S. offshoring took place. But not all the investments have been successful. China invested heavily in the manufacturing of solar panels with so far less economic success than it has had for steel, for example.

China was a poor agricultural society in 1950, and now it is a less poor and a manufacturing-oriented and increasingly urban society with pockets of great wealth in large urban centers. When I last visited China, I was certain that there were going to be changes in the goals and metrics used in the Chinese five-year plans. In June 2013, McKinsey & Company[22] noted that Chinese economic growth was slowing, wages and other costs factors were increasing, and supply chains were becoming more complicated. Chinese consumers were becoming more demanding, and the Chinese economy was losing competitiveness to Vietnam and others places. The report argues that China needs to improve the training of its managers to achieve greater efficiency, that is, engineering would not solve its problems. It also needs to build stronger research and development units, try to figure out how to deliver and receive components and final products on schedule. Eloot et al.[22] note that that manufacturing's direct contribution to a national economy peaks when it reaches 20% to 35% of GDP. When this chapter was written China's was 40% of the national GDP. The 12th five-year economic plan by the government for fiscal 2011–2015 identified biotechnology, information technology, new energy, environmental maintenance, new materials, high-end manufacturing and alternative fuels as priorities.

Reuters[23] reported in January 2016 that falling prices and overcapacity in the steel and energy industries were causing the Chinese manufacturing sector to shrink and more of the economic burden was on the service industry. A few months later, Gough[24] noted that China's economy was slowing to 6.7%, a rate of increase that would thrill many developing counties. Mangneir[25] observed that the Chinese government portrayed the news as part of an effort to rebalance the economy by slowing industrial production. I had a discussion about this same subject and reached the same conclusion on my last trip to China.

Whatever the stated and unstated motives, both of the two top dogs in world manufacturing have been heavily scrutinized by the public, the media and those in power. There is evidence that both as well as other nations are not abandoning manufacturing, and hence locations are needed for manufacturing. More specifically, we will need more facilities to produce steel, paper, furniture, refrigerators, and places to manage waste products and generate electricity. In addition, we know that nanotechnology, artificial intelligence, robotics 3-D printers, biotechnology, and other new manufacturing production lines and waste management facilities will be necessary for the emerging products, How many of these do we need and where will they go are the questions? Many of the industries will be noxious and need to be accommodated, as they will in Nigeria, Vietnam, and other hot zones, wherever they may be. And the cold zones will continue to generate waste products that need to be managed.

In 1965, recognizing that economic hot zones were in the United States, Japan and parts of Europe, I would have felt confident that I knew what methods to use to find the best sites. A half century later, I would face many more issues and be much less certain that I had all the important information. In the end, decisions have to be made, and this book focuses on processes, criteria and tools for informing those decisions.

A working definition of noxious facilities

My starting definition of a noxious facility noted in the preface was that it emits contaminants into the air and water, is unsightly, and sometimes can cause health impacts. In this section, I briefly review some alternative working definitions.

A consensus definition of a noxious facility from paper and web sources is that it is a location that produces and emits substances that are harmful to human physical and emotional health, which is a good but vague first step. I consulted four different sources to gain clarity:

- City planning literature;
- Government environmental protection databases;
- Census of business and industry databases; and
- Public perception literature.

City planners

Zoning and building codes are legal statements of land uses and activities allowed by local government in specific places. Zoning focuses on land use types, and building codes on how land uses are to be placed and built in order to protect human health and safety. Zoning normally separates noxious industry from residential, commercial, and non-noxious manufacturing facilities. However, as my preface example of the noxious leather clothes facility showed, normally does not mean always.

CityScope Town Planners, a company of town planning and property development consultants based in Pretoria, South Africa, defines and lists noxious industry.[26,27] Paraphrasing, they attach the label noxious to an industry that produces physically or mentally harmful gases, smells, noise, dust, smoke, etc., and thereby is excluded from normal industrial zones.

Supporting their definition is a list of approximately 70 activities that are labeled noxious industry. I have listed the first two and last two below:

- Blood boiling;
- Tallow melting;
- Works for the production of amyl acetate, aromatic esters, butyric acid, caramel enameled wire, glass, hexamine, iodoform, lampblack, B-Naphthol, resin products, salicylic acid, sulphonated organic compounds, sulfur dyes, ultramarine, zinc chloride, zinc oxide;
- All refining and works dealing with the processing or refining of petrol or oil or their products.

Their list includes industries that produce organic and inorganic chemicals, metals, petroleum, cleansing and other products. Notably, however, the list does not include electrical power generation and waste management, among others that I would expect to find on a list of noxious industrial sites.

Government environmental databases

Some governments collect data from facilities that emit and manage toxic substances. In the United States, the Toxic Release Inventory (TRI) database was created in 1986 and the data are publicly available.[28,29] The database reports toxic chemical releases and waste management activities, such as recycling and treatment. Each year, chemical, gas, oil, paper, and mining facilities that produce 25,000+ pounds, or 10,000+ pounds of chemicals listed as hazardous by the U.S. Environmental Protection Agency (EPA) must report to the TRI database.

However, the TRI database should not be taken at face value as defining a noxious facility. Being part of TRI does not mean that anyone has been exposed at a toxic level, or that its emissions smell and pollute the surrounding area. Equally important is the reality that not being in the TRI database does not mean that there is no risk from facilities. A relative small emission that is not part of the TRI database but is located in a dense urban area where the air sometimes is trapped, or a discharge into ground water in a rural area where people drink the water can be hazardous to human health and safety. A much larger emitter in an area with few people to expose, where the wind blows emissions away, and the people drink water that comes from many miles away may be less of a threat, assuming there is no acute event.

I have conversations every year with reporters and residents who have looked at TRI data and decided that an industrial facility has caused cancer or some other cluster of diseases. Invariably, they use TRI data to justify their conclusion. The real world is a lot more complicated than the statement that every nearby noxious facility causes cancer or some other illness. Someone has to be exposed before a cause-and-effect can be justified, and then other explanations of a time–space cluster must be explored, if indeed there is one.

A second major issue in regard to defining noxious using TRI is that not everyone is reporting all their information to the TRI database. Over 50 other countries have adopted TRI-like databases[29] and so it is possible to compare and contrast countries and find trends. For example, in 2003 about 25,000 facilities reported data to the U.S. TRI, and in 2014, the number was 21,783.

EPA summary reports of the TRI data help identify the magnitude of different industrial contributions to noxious emissions. Using the three- and four-digit North American Industry Classification System (NAICS, see below for a presentation), EPA grouped its reports into 27 business sectors. The production-related waste database shows the following division by sector in pounds of waste in 2014:[28]

- Chemicals – 40%;
- Primary metals – 11%;
- Petroleum – 9%;
- Metal mining – 7%;
- Electric utilities – 7%;
- Paper plants – 6%;
- Food/beverage/tobacco facilities – 6%; and
- All others – 14%.

EPA's reports also allow us to view the geographical distribution of emissions, and recycling. For example, the chemical industry data (3460 facilities) shows concentrations in the Northeast corridor, the Gulf coast, the Great Lakes region, and the California coast. In contrast, metal mining emissions (89 facilities), which are enormous in volume of emissions and increasing, are located in western states that mine gold, copper, silver, and gold with outliers in Missouri, Tennessee, and Alaska where zinc and lead are mined.

The map of 573 electric generating utility facilities resembles a map of the concentration of population in the United States. Note, however, this list reports fossil fuel facilities, not nuclear ones, which are not part of the TRI database.

Users can access historical TRI data. For example, the reports for the electric utilities show a 75% decrease in on-site air releases, by far the largest (see pollution prevention, Chapter 5).

The last industry sector to be briefly described here is federal facilities. Beginning with an order signed by President Clinton in 1993, federal facilities are required to report to the TRI database. In 2014, 446 did; almost two-thirds were U.S. military facilities. This list also includes federal prisons, police firing ranges, and government fossil fuel electric power facilities.

The reader can also obtain state and local reports. I checked the TRI database for the location of New Brunswick, NJ, where I work. A total of 25 facilities were reported within 2.7 miles of my office. Most were chemical plants, but the university where I work, the two major hospitals and a gas plant were on the list. Notably, the notes provided about each indicate that only four of the 25 even reported to TRI. In other words, only four had high enough emissions to report. The others report their activities, and the TRI database tells the reader based on the facility permits whether it could emit to the land, air and/or water, and if the information from the plant is up to date.

EPA's relatively new *EJScreen* database (type in EJScreen) allows readers to place an electronic pin or polygon on a map and learn not only the demographic attributes of the people who live in the area but also information about TRI sites and various other indicators of potential environmental exposure. As part of the presentation in Chapter 3 and elsewhere in the analyses for this book, I used *EJScreen* and other databases to learn more about communities where potentially noxious facilities are located.

Census of business and industry databases

Selected manufacturing and waste management facilities are the focus of this book. Mining and agriculture can be noxious, but their locations are largely resource determined, and consequently their location challenges are beyond the scope of this book. Hay's[30] and Visser's[31] papers provide location theory for these economic sectors that approximate the period reviewed in Chapter 2.

Manufacturing is assembling materials and producing new products by using biological, chemical, and physical processes. By the mid-twentieth century, business classification systems were developed to record trade flows, estimate economic impacts, and understand the lifecycle of resources and manufactured products. Business sector classifications are important, and concerted efforts have been made to standardize classifications to facilitate international trade. I briefly review four of these.

International Standard Industrial Classification (ISIC)

First adopted in 1948 by the United Nations, ISIC has focused on standardizing or translating individual national codes to a common one. The limitation of the system is that it does not go into the depth of several of the others. However, it is more than adequate for comparing major industrial sectors.[32–36] In essence, ISIC is the world gold standard.

Nomenclature of Economic Activities (NACE)

NACE is the classification system developed by European nations to facilitate intra-European and international trade. NACE is exactly the same as ISIC at the highest aggregation levels.[32–36]

Standard Industrial Classification (SIC)

SIC codes were adopted by the United States in 1937 and used to classify industry. They were replaced by the NAICS codes (see below). I mention it here because many manufacturing location facility studies in the United States described in Chapter 2 were based on the SIC system.[35,36] The SIC codes were 20–39, and easy to use, for example chemicals were SIC 28, petroleum refining SIC 29, primary metal SIC 33, and so on.

North American Industry Classification System (NAICS)

I reserved NAICS for last because as a U.S. resident I have often used it. It was constructed to harmonize data collected by Canada, Mexico and the United States. NAICS classified on the basis of production. The structure is different from ISIC and NACE, but is comparable to both at the 2-digit level, and, furthermore, experts have built translation codes that allow comparisons.[35-37]

NAICS goes down to 5 or 6 digits. I illustrate with pet food.

2-digit: major business group (31–33 is manufacturing)
3-digit: subsector (311 is food manufacturing)
4-digit: industry group (3111 is animal food manufacturing)
5-digit: NAICS industry (31111 is dog and cat food manufacturing)
6-digit: National industry (311111) is U.S. dog and cat food manufacturing

Using NAICs I can find out where in the U.S. major producers are located, and the NAICS site directs me to the 4-digit SIC code (2047 – dog and cat food). If dog and cat food is a noxious industry, then I can find the locations of those that are listed in the TRI database.

The waste management industry is a little more difficult to work with. NAICS places it in a 3-digit code 562, which includes:[38]

5621 Waste collection
5622 Waste treatment and disposal
5629 Remediation and other waste management services

They report 27,000 businesses and a workforce of 400,000 employees. Even with the NAICS codebook, however, I find it difficult to initially find the proper code for each facility type.

Overall, the NAICS codes work satisfactorily for some of the waste management sector, that is, the vast majority of these facilities produce odors, dust, noise and otherwise distress people, including trucks hauling waste through neighborhoods. However, NAICS codes are too broad to allow separation of noxious from other manufacturing industry. Hence, NAICS by itself is ineffective to identify noxious manufacturing facilities.

Public perception

The identifier noxious facility is human-centric. The first zoning and building codes were created to protect people. A century later, the TRI database was created in response to widespread public concern about human health, especially environmental-caused cancer. These realities require that for purposes of this book, we examine public perception of noxious facilities. I begin with a well-known list of nine acronyms associated with noxious facilities. These are in alphabetical order.[39-53]

- BANANA – Build absolutely nothing anywhere near anyone;
- LULU – Locally unwanted land use;
- NAIMBY – Not always in my backyard;
- NIABY – Not in anyone's backyard;
- NIMBY – Not in my backyard;
- NIMTOO – Not in my term of office;
- TISE – Take it somewhere else;
- TOADS – Temporarily obsolete abandoned derelict sites;
- WESER – Waste and source reduction.

Each of these has been applied to noxious facilities, although the last is not normally part of this list but needs to be part of the solution (see Chapter 5). Several of these acronyms also have been applied to facilities that many of us would welcome in our area. I would welcome a movie theater in my immediate neighborhood, a library, and a park. Yet some people do not, viewing these as source of unwelcome people who make noise and park in the street in front of their homes. Hence, these acronyms, while in use, have become overused and trite because every facility is probably noxious to someone.

I turn to random sample survey data. Fortunately, studies go back many decades. One of the most useful ones, published in 1980, compared the willingness of people in the United States to live near selected facilities. Mitchell[39] found the following proportion were willing to live within 10 miles of these land uses:

- Nuclear power plant and chemical hazardous waste disposal facility – 25%;
- Coal fired power plant or large factory – 70%;
- 10-story office building – 90%.

In 1983, Lindell and Earle[40] published a study of perceptions of industrial facility types. Their results are summarized here as willingness to live within ten miles of the following land uses:

- Nuclear waste disposal facility – 25%;
- Toxic chemical disposal facility – 30%;
- Nuclear power plant – 36%;

- Liquefied natural gas storage plant – 49%;
- Oil refinery: liquefied – 52%;
- Coal power plant – 59%;
- Oil power plant – 67%;
- Natural gas power plant – 78%.

These and other studies suggest levels of perception of noxiousness. I have conducted several studies similar to these, but in terms of the concept of noxious facility, a study conducted in 2010 told me more about the concept than the others because we asked people to react to the questions with images, colors and other perceptions rather than only with numerical ratings.[53] The second goal was to determine what risk management practices they prefer for nearby facilities. For example, we asked people to "think of a power plant, industrial site, or waste management site in your area." If they named a site, then we asked them to tell us what it produces or does. If they could not think of one, we asked them to think of the one closest that they could identify. A key point was that we asked for only one site because we wanted to focus on the site that was foremost in their mind.

For those who could identify a site we asked them to close their eyes and state one feeling, emotion, image and color that they attach to the facility. All of this information was recorded verbatim. By asking for one site, one feeling, one emotion, one image, and one color we obviously limited the response to the first site in their mind, which was what we wanted in order to tap into raw emotions.

A total of 651 valid responses were collected, focusing on the following facilities:

- Energy – 32.0%;
- Waste management – 25.8%;
- Manufacturing – 7.4%;
- No site mentioned – 34.9%.

I call your attention to the fact that energy and waste management facilities attracted more responses than did manufacturing ones, despite the reality that there are many more manufacturing sites than waste management and energy ones. Within the set of identified site types, only two types were identified more than 50 times: 64 identified nuclear power plants and 55 landfills. However, identification did not necessarily mean negative identification. In fact, 48.5% of respondents did not identify even a single feeling-emotion about the site they named. Six percent of those who did list one expressed feelings such as, "indifference," "confusion," "curiosity," but neither a positive or negative feeling-emotion.

Only 45.5% offered a positive or negative comment about the land use they chose. Of these, 26.1% offered a negative comment. "Fear," "concern," "worry" and "dangerous" were mentioned at least ten times, and 19.4% offered a positive one such, such as "safe," "secure," and "well managed." Mixed reactions were also asserted by respondents about images and colors. Space limitations do not permit a full recitation of the results. A few are highlighted because they say a lot about why some facilities are more noxious than others.

Only ten people identified nuclear waste facilities, but the feelings/emotions that they disproportionately attached to the facility were "bad," the image was "explode," and the colors were "red" and "yellow." Red, yellow and orange are aggressive colors, considered a sign of fear and concern.[54-56] A total of 64 identified a nuclear power plant. "Bad" was associated with this land use, but so was "needed." The color "red" was disproportionately indicated, but so was the color "blue," among the most positive colors in the context of risk. The word "smoke-stack" was disproportionately listed and yet so was the word "clean." Clearly, some people valued and had positive images of nuclear power plants, not so of nuclear waste management facilities in this small sample of United States residents.

Landfills and other waste management facilities were associated with the words "stink," "ugly," and "brown" and electricity-generating stations using coal were associated with the colors "black" and "gray."

The second part of this survey was to ask people to indicate their response to nine risk management options ranging from forbidding access to the site to regularly monitoring the health who live the near the site. The higher the number, the more the respondents wanted strong risk management for the sites. With a scale ranging from 1 to 10, where 1 meant a very low priority and 10 a very high priority, 4 of the 10 land uses stood out because they had average scores of 8+ on nearly all of the nine management options. These four land uses were as follows:

- Nuclear waste management facilities – 7;
- Chemical plants – 7;
- Nuclear power plant – 5;
- Oil refinery – 5.

Despite using a survey method different from other studies, the results of this one were similar to those conducted three decades earlier with more conventional survey questions. Specifically, six of the titles used in the 1983 survey described above were the same as the 2010 one. The Spearman rank correlation between the two sets of results was 0.77 ($p = 0.007$). The emotions-feelings, images and colors attached to these land uses and the distance metric used by Lindell and Earle[40] found the same set of most enduringly objectionable noxious facilities.

A working list of noxious sites for an urban environment

If you can smell it, hear it, see it or it is perceived as or is a threat to your health, your friends and neighbors, the vast majority of us are likely to consider a facility to be noxious. Among the nine listed below, the first three are the most likely to be considered noxious, although the distress associated with nuclear power plants is partly balanced by the feeling that some have that nuclear power represents clean energy.

- Waste management, nuclear, chemical and conventional waste;
- Chemical plants;
- Nuclear power plants;
- Oil refining and gas plants;
- Metal smelting and fabricating plants;
- Coal-fired electricity-generating plants;
- Building product manufacturing;
- Paper plants;
- Animal products – leather tanning, slaughter houses.

This list of nine does not include every noxious facility nor is every facility on this list noxious. Two years ago, I visited a new TRI-list facility that I cannot name. The inside of the plant was so clean, I joked that I could have eaten my lunch while sitting on the floor. In contrast, in 2014, Irwindale, California, sued the Huy Fong foods company after the company spent $50 million on a 650,000 square foot factory that produced sriracha chili sauce. In the manufacturing process, the plant produced a combination of garlic and chili that drifted over part of the town. The town dropped the suit after Huy Fong upgraded the filtration system, a clear application of pollution prevention.[57]

Outline of the book

With a backdrop of remarkably uneven and rapid world economic growth and identification of a set of focal noxious facility types, the remainder of this chapter summarizes Chapters 2–10. These nine chapters focus on how noxious facilities were sited during much of the last century, why those approaches are no longer adequate, what new processes and tools are now essential, and how analysts can keep up with this evolving subject.

The book is divided into three parts. Part I has two chapters. Chapter 2 describes and explains traditional location theory focusing on manufacturing and shows how manufacturing became the epitome of the desired community asset. Today, this would be considered to be "old" location economics and some would dismiss it. But in addition to the fact that the siting of noxious industry was a key part of urban industrialization, the reality is that location economics remains critical, but changed. Chapter 3 uses six small communities located within 12 miles of the author's home to illustrate how industry, especially what would now be called noxious industry, became the growth engine and what happened to these places as the world and rules changed, leaving these places with undesirable industrial and waste management legacies.

Part II consists of five chapters that collectively illustrate what noxious manufacturers and waste managers have been doing to cope with greater uncertainty that characterizes the early twenty-first century. Chapter 4 explains how the traditional location factors have changed and how new factors must be addressed, including some that are rapidly evolving and others that have been prominent for decades. The chapter asserts that risk analysis when combined with location economics can serve as a framework for better understanding and managing added uncertainty. However, both need to be better able to deal with the greater complexity of multiple parties, varying environments, and life cycle challenges than they have in the past.

Chapters 5 to 8 offer a set of siting and non-siting options that are separated into individual chapters. Each of these case study chapters focuses on a process tool, shows how it can serve to reduce human health and safety risks and decrease the chances that an investment will be an economic failure.

Chapter 5, for example, is about becoming less noxious by practicing pollution prevention, recycling and choosing non-noxious producing options, or in other words, using inherently safer and cleaner designs and processes. A less noxious, safer and greener facility may be a more desirable neighbor that pays taxes, hires local people and does not pollute the local environment. However, a less noxious facility at one location could cause issues elsewhere; for example, recycling facilities have proved to be problematic in some locations. Merck is the major example and several international examples are more lightly treated.

Chapter 6 defines CLAMP, which means concentrating locations at major plants already controlled by the organization. Clamping should reduce the risk of antagonizing some officials and residents, while continuing investments in places where people want these facilities or will at least tolerate them. CLAMP is a relatively safer option in general. Yet a CLAMP policy may lead to social and environmental justice issues. If you hope to use the CLAMP policy, attention must be focused how to make it less noxious than the existing facilities. The effort by the U.S. Department of Energy to store elemental mercury is the main example, and others, several international, are more briefly discussed in this chapter and in Chapters 5 and 7 to 9.

Chapter 7 focuses on negotiating with government, local publics and the use of a variety of analytical and process approaches to find a location for facilities that might not otherwise be locatable in the desired area. Most of these efforts have not been successful when the proposed site is high on the noxious facility site, but there have been some successes. The chapter will illustrate both successes and failures. The two major examples focus on the Waste Isolation Pilot Plant (major success) and Yucca Mountain (major failure) targeted for nuclear waste storage. Briefly covered international examples are sites for nuclear waste from Finland, Sweden and Germany.

Chapter 8 surrenders to the reality that some government and private organizations may need to stop manufacturing part or all of a product, move manufacturing to another place and in other ways remove the production and waste management from its portfolio of activities in its current locations. These "letting it go" options have not always been successful, if by success one means a new industry that produces economic and environmental benefits results. A painful result is the growth of the lead battery industry in China, which has had challenging public health consequences.

Part III focuses on keeping up with the rapid pace of change. Chapter 9 briefly summarizes 24 social/political, economic/business, and environmental/ health tools that have been suggested and sometimes used as aids in siting decisions. In each case the objective of the tool, typical applications, prerequisites for use, assessment of its use in siting noxious facilities, and several citations are presented. This is followed by a list of five books that I consider classics for those that want to look back, a dozen web-based locations to look in order to keep up with trends driving the location of noxious facilities, and five U.S. government websites. I also show how I use Google Scholar to keep up with academic articles on the subject.

Chapter 10 asserts that developers and governments need to be prepared to cope with a seemingly continuously changing world economic environment. I begin with offering six ethical principles to guide noxious facility siting. Next, I argue that traditional economic and risk analysis approaches need to be supplemented to better define legitimate participants, to incorporate and display data that are more easily grasped by participants and guidelines to help them in their interactions and deliberations, and to create capacity to scan for trends that will impact the need for siting noxious facilities rather than passively wait and then react to a proposal. This chapter uses the chronic problem of finding siting and non-siting approaches for "the" most chronic noxious facility siting problem, which is managing garbage in the world's largest cities, for example, New York City and New Delhi. The chapter also addresses how to cope with surprises, such as when a government decides to open up previously unavailable sites to industrial development. These examples are used to illustrate the processes and tools presented in Chapters 5 to 9.

References

1 Delong. Estimates of World GDP, One Million B.C.–Present: My Views as of 1998. 1998. Delong.typepad.com/delong_long_form/2014/05/estimates-of-world-gdp-one-million-bc-present-1998-my-view-as-of-1998-the-honest-broker-for-the-week-of-may-24–2014.htm. Accessed January 10, 2017.

2 The Conference Board, Global Economic Outlook 2017 – Charts and Tables. www.conference-board-org/data/globaloutlook/index.cfm?id=27451. Accessed January 10, 2017.

3 Defense Security Service. Targeting U.S. Technologies: A Trend Analysis of Cleared Industrial Reporting. August 12, 2016. www.dss.mil/documens/ci/16–08–15_unclass_trends_with_cover.pdf. Accessed August 18, 2016.

4 Collier P. *The Bottom Billion*. New York, Oxford University Press, 2007.

5 World Bank, 2014. Table 4.2. Economy. World Development Indicators: Structure of Output. www.worldbank.org/table/4.2# www.worldbank.4.2_Structure–of-Output.pdf. Accessed October 3, 2016.

6 Central Intelligence Agency. *The World Fact Book*. www.cia.gov/library/publications/the-world-factctbook/geos/xx.html. Accessed October 3, 2016.

7 MarketsandMarkets Research Private Limited. Industrial Waste Management Market Worth $1,442.0 Billion by 2019. www.marketsandmarkets.com/PressReleases/Industrial-waste-management.asp. Accessed November 18, 2016.

8 Global Marco Monitor. Chart of the Day: US Manufacturing Employment, 1960–2012. 2012. https://pro.creditwritedowns.com/2012/05/chart-of-the-day-us-manufacturing-unemployment-1960-2012.html. Accessed March 27, 2018.

9 Ebinger C, Massy K, Avasarala G. Liquid Markets: Assessing the Case for U.S. Exports of Liquefied Natural Gas. *The Brookings Institute. Policy Brief 12–01*, May 2012. www.brookings.edu/research/liquid-markets-assessing-the-case-for-u-s-exports-of-liquefied-natural-gas/. Accessed March 27, 2018.

10 U.S. Energy Information Administration. *Effect of Increased Natural Gas Exports on Domestic Energy Markets*. http://energy.gov/sites/prod/files/2013/04/fo/fe_eia./ng.pdf. Accessed May 12, 2016

11 Montgomery D, Baron R, Bernstein P, Tuladhar S, Xong S, Yuan M. Macroeconomic Impacts of LNG Exports from the United States. www.nera.com/publications/archive/2013/macroeconomic-impacts-of-long-experts-from the United States.html. Accessed May 16, 2016

12 Deloitte, Made in America: The Economic Impact of LNG Exports from the United States. www.deloitte.com/view/en_US/us/industries/oil-gas/9f70dd1cc9324310VgnVCM1000001a56f00aRCRD.htm. Accessed May 15, 2016.

13 Treyz F, Brooks R, Nystrom S, King B, Cook C, Morton C. *The Macroeconomic Impact of LNG Exports: Integrating the GPEM^R Natural Gas Model and the PI^{+R} Regional Model*. Washington DC, REMI, Inc., 2015. www.usaee.org/usaee2012/submissions/onlineproceedings/RBAC%20REMI%20LNG.pdf. Accessed May 16, 2016.

14 Scott R. *The Manufacturing Footprint and the Importance of U.S. Manufacturing Jobs*. 2015. www.epi.org/publication/the-manufacturing-footprint-and-the-importance-of-u-s-manufacturing-jobs/. Accessed March 27, 2018.

15 Rothfeder J. Why Donald Trump is Wrong about Manufacturing Jobs and China. *The New Yorker*. March 14, 2016. www.newyorker.com/business/currency/why-donald-trump-is-wrong-about-manufacturing-jobs-and-china. Accessed March 27, 2018.

16 Immelt J. The CEO of General Electric on Sparking an American Manufacturing Renewal. *Harvard Business Review*. March 2012. https://hbr.org/2012/03/the-ceo-of-general-electric-on-sparking-an-American-manufacturing-renewal. Accessed March 27, 2018.

17 Denning S. Decoding Jeff Immelt and GE's New Act: U.S. Manufacturer. *Forbes*. 2013. www.forbes.com/sites/stevedenning/2013/02/11/decoding-jeff-immelt-ges-new-act-us-manufacturer/#637ce1594196. Accessed March 27, 2018.

18 Nisen M. CEO Jeff Immelt: Here's the Case for Making Things in American Again. *Business Insider*. 2013. www.businessinsider.com/jeff-immelt-heres-the-case-for-American-manufacturing-2013-2. Accessed March 27, 2018.

19 Blodget H. CEO Jeff Immelt on Transforming GE – Reflections on Winning the Race, Digitizing Manufacturing and Leading 311,000 People Into a new Age. *Business Insider*. 2015. www.businessinsider.com/interview-with-ge-ceo-jeff-immelt-on-transforming-ge-2015-12. Accessed March 27, 2018.

20 Caro R. *The Power Broker*. New York, Alfred Knopf, 1974.

21 Bradsher K. 2012. Social Risk Test Ordered by China for Big Projects. *New York Times*, www.nytimes.com/2012/11/13/world/asia/china-mandates-social-risk-reviews-for-big-projects.html. Accessed November 17, 2016.

22 Eloot K, Huang A, Lehnich M. A New Era for Manufacturing in China. 2013. www.mckinsey.com/business-functions/operations/our-insights/a-new-era-for-manufacturing-in-china. Accessed March 27, 2018.

23 Reuters. China Manufacturing Sector Shrinks at Fastest Rate for more than Three Years. 2016. www.theguardian.com/business/2016/feb/01/china-manufacturing-sector-shrinks-for-sixth-straight-month-january-figures-show. Accessed March 27, 2018.

24 Gough N. What China's Economic Growth Figures Mean. *New York Times.* July 14, 2016. www.nytimes.com/2016/07/15/business/international/china-gdp-economic-growth.html. Accessed March 27, 2018.

25 Magneir M. As Growth Slows, China Highlights Transition from Manufacturing to Service. *Wall Street Journal.* 2016. www.wsj.com/articles/as-growth-slows-china-highlights-transition-from-manufacturing-to-service-1453221751. Accessed March 27, 2018.

26 CityScope Town Planners. Town Planning Terminology Explained. http://cityscope.co.za/town-planning-terminology-explained. Accessed November 1, 2016.

27 CityScope Town Planning Definitions, SANS 10400, Building Regulations, sans10400.co.za/town-planning-definitions/. Accessed November 18, 2016.

28 U.S. Environmental Protection Agency. TRI National Analysis 2014, Industry Sectors, updated January 2016. www.epa.gov/sites/production/files/2016-01/documents/tri_na_2014_complete_english.pdf. Accessed March 27, 2018.

29 U.S. Environmental Protection Agency. TRI Around the World. www.epa.gov/tri nationalanalysis/tri-around-world. Accessed March 27, 2018.

30 Hay M. A Simple Location Theory for Mining Activity. *Geography.* 61, 2, 65–76, 1976.

31 Visser S. On Agricultural Location Theory. *Geographical Analysis.* 14, 2, 167–176, 1982.

32 Department of Economic and Social Affairs, Statistics Division, United Nations. *International Standard Industrial Classification of Economic Activities (ISIC) Revision 4,* New York, United Nations, 2008.

33 Nicita A, Olarreaga M. Trade, Production, and Protection Data Base, 1976–2004. *The World Bank Economic Review.* 21, 1, 165–171, 2007.

34 Remond-Tiedrez I, Defense-Polajarv P. Conversion of U.S. Supply and Use Tables Using the European Classification, First Analysis for the Years 2008 and 2009. Paper presented at the International Input-Output Association. Kitakyushu, Japan, July 2013.

35 Office of Management and Budget, Executive Office of the President. *North American Industry Classification System (NAICS).* Washington DC, OMB, 1999.

36 NAICS Association. 2016. NAICS Identification Tools. www.nacis.com/search/. Accessed March 27, 2018.

37 NAICS Association. Six Digit NAICS Codes and Titles. www.naics.com/six-digit-naics/?code=3133. Accessed August 29, 2016

38 U.S. Bureau of Labor Statistics. 2016. Waste Management and Remediation Services: NAICS 562 www.bls.gov/iag/tgs/iag562.htm. Accessed August 29, 2016.

39 Mitchell R. for the U.S. Council on Environmental Quality. 1980. *Public Opinion on Environmental Issues: Results of a National Public Opinion Survey.* Washington, DC, U.S. Governmental Printing Office, 1980.

40 Lindell M, Earle T. How Close is Close Enough: Public Perceptions of the Risk of Industrial Facilities. *Risk Analysis.* 3, 245–253, 1983.

41 Kemp R. Why Not in My Backyard? A Radical Interpretation of Public Opposition to the Deep Disposal of Radioactive Waste in the United Kingdom. *Environment and Planning A.* 22, 1239–1258, 1990.

42 King R, Mauer M, Huling T. An Analysis of the Economics of Prison Siting in Rural Communities. *Criminology & Public Policy.* 3, 3, 453–480, 2004.

43 Lapp R. One Answer to the Atomic Energy Puzzle—Put the Atomic Energy Power Plants in the Ocean. *New York Times Magazine,* June 4, 20–22, 80–84, 90, 1972.

44 Gerrard M. *Whose Backyard, Whose Risk.* Cambridge, MA, MIT Press, 1994.

45 Greenberg M, Popper F, West B. TOADS: A New American Urban Epidemic, *Urban Affairs Quarterly.* 25, 438–457, 1990.

46 O'Hare M, Bacow L, Sanderson D. *Facility Siting and Public Opposition.* New York, Van Nostrand and Reinhold, 1983.

47 Popper F. Siting LULUs. *Planning,* April (reprinted in American Planning Association, *The Best of Planning,* 1981 [Chicago, Planners Press, 1989]).

48 Portney K. *Siting Hazardous Waste Treatment Facilities: The NIMBY Syndrome.* New York, Auburn House, 1991.

49 Slovic P, Finucane M, Peters E, MacGregor D. Risk as Analysis and Risk as Feelings: Some Thoughts about Affect, Reason, Risk, and Rationality. *Risk Analysis.* 24 (2), 311–322, 2004.
50 Useem B. Prison Siting and Economic Development. *Criminology & Public Policy.* 3, 3, 451–452, 2004.
51 Wald M. NRG Abandons Project for 2 Reactors in Texas. *New York Times.* April 20, p. B6, 2011.
52 Marquart J. Economic Development Is in the Eyes of the Beholder. *Criminology & Public Policy.* 3, 3, 489–492, 2004.
53 Greenberg M, Popper F, Truelove H. Are LULUs Still Enduringly Objectionable? *Journal of Environmental Planning and Management,* 55, 6, 713–731, 2012.
54 Birren F. *Color Psychology and Color Therapy: A Factual Study of the Influence of Color on Human Life.* New Hyde Park, NY, University Books, Inc., 1961.
55 Color Wheel Pro. Color. 2010. www.color-wheel-rpo.com/color-meaning.html. Accessed November 18, 2010.
56 Kaya N and Epps H. Color-emotions Associations: Past Experience and Personal Preference. AIC Color and Paints, Interim Meeting of the International Color Association, 2004. Proceedings. www.fadu.uba.ar/sitios/sicyt/color/aic2004/03–34.pdf. Accessed December 19, 2010.
57 Feldman D. Southern California City Drops Bill Against Sriracha Hot Sauce Plant. May 29,2014.www.reuters.com/article/us-usa-sriracha-california-idUSKBN0E90AU20140529. Accessed March 16, 2018.

Part I

Industrial location theory and practice in the twentieth century

2 Urbanization, industrialization, and noxious facilities

Introduction

This chapter has three goals:

- Describe the relationship that was assumed to exist between manufacturing and the growth of cities in the nineteenth and much of the twentieth centuries;
- Review location considerations for noxious industry and waste management facilities during this period; and
- Summarize the evolution of traditional economic location theory and tools, especially those elements that bear upon industrial and waste management facilities.

Cities and manufacturing

Industrial location theory is historically important because of its relationship to the growth of cities in the United States, Europe, Canada and many other locations, and in turn the raising of quality of life in these areas. The literature distinguishes between a "basic" industry, one that produces for customers outside the local area and a non-basic industry for internal consumption.[1-8] This distinction is critical because economists, economic geographers, and regional scientists, as well as elected officials assumed that a city would not thrive if people merely washed each other's laundry. There needs to be, they argued, economic activities that bring money into cities from outside sources. This external money would grow and sustain the city. That source of the external money was almost always identified with manufacturing. In my undergraduate economic history course, for example, the professor built the entire course around this idea to the point that it seemed like she was teaching us a set of religious beliefs. The fact that she was teaching this course in a university in New York City, a massive industrial center at that time, I suspect, increased her fervent belief in manufacturing as the basis of growth and an increased quality of life.

Part of the belief system she taught was that a settlement begins on the basis of trade and commerce. Eventually, it gets a factory, produces goods sold elsewhere, and then proceeds into the feedback loop as described by Pred (see Figure 1.2) where more manufacturing leads to more innovations, more growth and trade, and more population.[1-3]

This following simple equation defines the basic–nonbasic dichotomy:

Total local product = basic economic product (B) + nonbasic economic product (N).

The B/N ratio was widely used with the understanding that basic industry drives growth. If half the population is defined as basic and half as non-basic, their ratio is 1:1. The assumption is that each basic job creates another non-basic one in this simple example.

Several analysts acknowledged that some basic activity served local markets, and they tried to calculate that number. Alexanderson[5] and Morrissett[9] defined a lower level of business sector activity that was to be found in every city. Morrissett's "k" value was the lower 5th percentile of each business sector as required of every city as a minimum. In regard to wholesaling, for example, Richland, Washington, and Oak Ridge, Tennessee, had the lowest proportion of wholesale workers of any places in the United States. This is not a surprising finding because both locations were major sites for the building of nuclear weapons and there was little of private capital investment in these locations.

Table 2.1 shows selected data from Morrissett's 1958 paper.

The data show that durable manufacturing; nondurable manufacturing (excluding food and printing and publishing); and the chemical industry along with mining have much larger non-local proportions of national jobs than they do "k" values, that is, they serve mostly non-local markets. Hence, they are basic-oriented industries. In other words, these sectors bring the largest relative share of outside dollars.

Table 2.1 Selected "K" values from Morrissett study of U.S. cities*

Industry grouping	"k" value, %	National value, %
Mining	0	0.9
Construction	3.5	6.2
Durable manufacturing	0.3	15.9
Nondurable manufacturing	1.6	14.2
(Food and related)	(0.7)	(3.0)
(Printing and publishing)	(0.7)	(2.1)
(All other nondurable manufacturing, including chemicals)	(0.2)	(9.1)
Transportation and utilities	2.9	9.2
Trade	14.2	22.6
Services	15.2	30.8
Total, all sectors	37.7	100.0

Note
* Adapted from Morrissett 1958, table 1.[9]

Various challenges were aimed at this basic–nonbasic distinction, including reliance on employment data to measure it, and the reality that some proportion of basic industry is required to support non-basic industry and vice versa. Furthermore, some cities have large white collar job bases that act as basic industry for their community.[10,11] Nevertheless, the literature of the 1950–1970s and policy decisions emphasized the importance of the growth of manufacturing industries. Many local government officials tried to recruit manufacturing. The reliance on manufacturing to grow the economy continues over a half century later in China, Vietnam, and other countries (see Chapters 1 and 8).

Cities were classified into ten types based on their employment attributes. Using 1950 census data, Chauncey Harris[12] classified 43% of 605 U.S. cities as manufacturing, by far the largest grouping among cities. My fellow graduate students and I spent hours studying Harris's methods and made our own maps of his classification by pasting a large piece of tracing paper over a map of the United States. Using the same data, but a different classification method, Nelson[13] developed a method that allowed multiple labels to be applied to cities. Nelson classified 27% as "diversified" followed by 20% as "manufacturing." This manufacturing city system classification evolved through various iterations. Duncan et al.[14] produced a refined classification that categorized the largest cities as possessing large industrial, financial, retailing and other sectors, whereas smaller cities specialized in manufacturing and other business sectors. While Nelson, Duncan and others took into account the reality that larger cities would be more economically diversified, they did not neglect the belief that manufacturing was the key city-builder.[15-18] In my generation, it would have been considered heresy not to accept that belief as reality.

Based on widespread acknowledgement of the importance of manufacturing not only by students and professors but also by people with money to invest and elected officials who wanted them to invest in their jurisdictions, it follows that cities would do everything that they could to satisfy the needs of manufacturers. But actually that was not always the case. Writing in 1954, Muncy[19] observed that courts were not preserving industrial-zoned land and were allowing some of it to be used for residential because more profit could be made from residential uses. He labeled industrial land as "zoning's stepchild," that is, it does not get fair treatment. Zoning officials, he asserted, assigned poor swampy and other bad lands that are flooded to industry.[19] Furthermore, residential construction, he observed, is allowed in industrial areas, and all sorts of other land uses are allowed in industrial zones, including junk yards, bars, loan companies, gas stations, and other undesirable land uses. Muncy pointed out that industry needs space for parking and loading and off-loading and these "undesirable" land uses interfere with industry. In short, he says that "heavy" industry is excluded or given the poorest land in poor neighborhoods. What is so fascinating about Muncy's work is that he considers so-called "heavy industry" less noxious than commercial uses; the latter, he argues, create more noise. He observed that industry is needed to grow the economic base and making it difficult for industry has social, economic and fiscal costs.

Muncy, writing in 1954, suggests the following policy reforms to support manufacturing (see also Wrigley for an illustration[20]):

- Residential uses should not be allowed in areas zoned for industry;
- Performance standards for industry should be substituted for obsolete industrial zoning classifications; and
- Special industrial districts should be organized to improve efficiency and compatibly with nearby residential districts.

In his well-known 1960s text book *The American City*, Raymond Murphy[21] states that about 5% of land use of cities is devoted to industrial facilities. This was a central tendency and some devoted much more than 10% of their land to industrial plants, not including infrastructure (see Chapter 3 for some illustrations). When Murphy's text was published in 1966, industrial location was dispersing, a pattern that has persisted for more than a half century. Within the United States, the textile industry and then many other industries moved from the Northeast to the South.[22] Thereafter, many began to move out of the United States to Mexico and Asia. The Northeast and Midwest movement to the South was underway when I was first studying industrial location theory in the 1960s, and several of my professors and student colleagues thought that these shifts would be reversed. They markedly increased (see Chapters 1 and 8 for more discussion). Beliefs, perhaps faith is a better word, did not quickly die about a great deal of location theory conventional wisdom.

My professors were less skeptical about what seemed to be a city to suburban movement. Murphy,[21] for example, noted that most cities had areas set aside for industry, usually adjacent to waterways and rail lines in river valleys. This land, he noted, was where steel, petroleum and other so-called "heavy" industries located. Suburban location was only possible in some locations, or so was the consensus well into the twentieth century.

While industry was desirable, some analysts were concerned about environmental impacts. For example, Carey and Greenberg,[23] both born in and familiar with industrial neighborhoods in New York City, examined the New York metropolitan region for these edge city industrial locations, observing that sewage plants and other types of noxious facilities were disproportionately located at edges, and the fumes from these wafted over adjacent suburbs, and vice versa. They argued that edge locations for noxious industries made perfect sense. Cities could continue to obtain revenues and diffuse some of noxious emissions.

One of the most interesting historical studies of mid-twentieth century suburbanization of industry was by Kitagawa and Bogue[24] who examined intraregional trends in decentralization of U.S. industry. They noticed suburbanization of manufacturing first during the period 1929–1939 but they suggested that it seemed to stabilize during the period 1939–1947 (a period marked by World War II).

Most notable was their list of factors involved in pushing manufacturing to what they called "ring" areas. I paraphrased and shortened their list of factors pushing industry out of cities to suburban rings:

- Only limited space in the central city is available because of unsuitable terrain, lack of undeveloped space, and inability to annex adjacent land;
- Modern factories need space for continuous processing on a single floor;
- Noxious industries are not suitable in residential and most commercial areas;
- Modern transportation does not funnel industry into central cities, rather it spreads it to adjacent areas and adjacent areas become accessible due to the building of the national highway system and extension of railroad spurs in those ring areas;
- Suburbs provide tax benefits, favorable zoning and building codes, and land to industry; and
- Hard working and reliable labor are available in suburbs.

They also listed factors that would keep manufacturing in large cities:

- Original central city locations may be the only place that has the accessibility to ports, flat and otherwise suitable land;
- The city has the legal ability to annex adjacent areas;
- Some industries may be able to reuse old multi-story buildings;
- Noxious industries have well-established zones, supported by zoning and building codes, and noxious industries are not wanted by adjacent areas;
- Water, sewers, electric power, natural gas, and other utilities may not be available outside the center city;
- Cities may work hard to maintain these industrial areas, especially when they are essential ingredients of other city businesses;
- Transportation access outside the city is limited;
- Cities offer financial benefits to keep manufacturing; and
- Large pools of skilled and unskilled labor are available in cities.

Overall, while scholars were writing about the key role of manufacturing in building urban centers, manufacturers were beginning to move from cities in the American Manufacturing Belt in the United States to the South and then West, outside the United States to Asia, and from historic manufacturing districts to suburbs. The cumulative impact of this change demonstrates the other side of building cities through manufacturing, that is, cities that lost their manufacturing competitiveness would shrink unless they could replace it with white collar business.

Table 2.2 compares the populations in 1910, 1960, and 2010 of ten key cities that formed the boundaries of the American Manufacturing Belt. The results are striking. These ten ranked 1–10 in population in the United States in 1910. The growth of New York, Chicago, Philadelphia, Cleveland, Baltimore, Buffalo and especially Detroit were spectacular between 1910 and 1960; each grew by more than 50%. A half century later, only New York City reported a larger population. Buffalo, Detroit, Pittsburgh, Baltimore, Cleveland and St. Louis experienced massive population losses. Four of the six were

Table 2.2 Population of ten populated American manufacturing belt cities: 1910, 1960, and 2010

City	Population 1910, millions, [rank]*	Population 1960, millions [rank]*	Population 2010, millions [rank]*	Harris classification of cities[12]
New York	4.769 [1]	7.782 [1]	8.175 [1]	diversified
Chicago	2.185 [2]	3.550 [2]	2.695 [3]	diversified
Philadelphia	1.549 [3]	2.002 [4]	1.526 [5]	manufacturing
St. Louis	0.687 [4]	0.750 [10]	0.319 [58]	diversified
Boston	0.671 [5]	0.697 [13]	0.618 [22]	diversified
Cleveland	0.561 [6]	0.876 [8]	0.397 [45]	manufacturing
Baltimore	0.558 [7]	0.939 [6]	0.621 [21]	diversified
Pittsburgh	0.534 [8]	0.604 [16]	0.306 [59]	manufacturing
Detroit	0.466 [9]	1.670 [5]	0.714 [18]	manufacturing
Buffalo	0.424 [10]	0.533 [20]	0.261 [70]	manufacturing

Note
* Rank in population among U.S. cities.

classified as manufacturing centers, and the other two were diversified, but with large manufacturing components.

Manufacturing losses were only part of the challenges that these cities faced. The growth of the U.S. highway system, the growth of suburbs aided by low interest loans, income and racial tensions in cities, the opening of interstate and ring highways, the ability of some Western and Southern cities to annex adjacent land, and changing urban-suburban politics, as well as selected other factors badly impacted many manufacturing-reliant cities.[25] However, the massive losses of population in these cities is at least partly explained by the decline in cities that had become manufacturing dependent. When I was a graduate student we were told that cities should emulate the Detroit model of growing based on the automobile industry. The New York City and Boston models seem to have been more sustainable.

Noxious industry needs

Whether the starting date was 1910 or 2010, the following ten are standard attributes on the list of assets that noxious manufacturing and waste management facilities need:

- Raw materials and energy;
- Labor and technology;
- Market and clients;
- Capital;
- Suitable land;
- Agreeable governments;
- Infrastructure, including water and waste management;

- Suitable local environment, including acceptable weather;
- Transportation;
- Entrepreneurship.

The following is a brief commentary about each of these ten.

Raw materials and energy

Producers and waste managers require raw materials. Steel, copper, aluminum, and other metal mills need ore or scrap. Paper mills need wood from forests or scrap paper. Landfills, incinerators and recycling facilities need waste. Nuclear power plants need enriched uranium, and coal, oil and gas plants need their organic fuels. Changes in price, environmental regulations, war, international and interstate trade arrangements, and other external factors can cause serious raw materials shortages or opportunities. Companies try to sign long-term deals to reduce this kind of uncertainty, but sometimes without success.

Noxious manufacturing and some waste management facilities can be voracious energy users. Many build their own power plants to take advantage of nearby energy sources. For example, the aluminum industry was a mainstay of the Washington economy relying on Columbia River hydropower. The U.S. Department of Energy's nuclear waste management facility in South Carolina relied on an old coal-burning power plant. Then an opportunity arose to close this eyesore and use wood scrap gathered from forest management work by the U.S. Forest Service on the site. The new wood burning plant is now operating. Some nuclear power plants have closed because shale gas is much cheaper. Chapter 3 illustrates how a mini-steel mill in New Jersey obtained inexpensive electric power to site in the state.

Producers have learned that their business is vulnerable to political actions that include arbitrarily changing the price and even cutting off energy supplies. This concern is not new. Writing decades ago, Alonso,[26] Stafford,[27] and Smith[28] pointed out that is industry should avoid what they called "peripheral" locations and nations because their resource needs are not guaranteed.

Labor and technology

A half century ago, in the U.S. and other developed nations, factories depended on skilled workers to operate expensive and dangerous equipment. Much of the U.S. industrial labor force belonged to unions, was paid relatively high wages, and had good health and other benefit packages. Workers who were loyal to their companies flourished in the United States and Western Europe for many years (see Chapter 3 for several examples).

However, even in 1960, the trend toward relocation based on cheaper, non-union workers and technology substituting for production line workers was underway (see Chapters 1 and 4). The evolution of the role of labor in manufacturing continues. Japan, Korea, China, and Vietnam have become major manufacturing nations (Chapter 1, Figure 1.1). Have the United States and Europe lost their advantages because of high labor costs? The answer is yes for some kinds of production (see Chapter 1) but not for others, and this evolution will continue.

In regard to noxious industries, technology has replaced workers and doubtless saved workers from fatal exposures. Chapter 3 discusses the case of Manville, New Jersey, which likely is the poster child for these cases.

An important point about workers at noxious facilities is that many of them are among the most well trained and disciplined employees. Unlike the days when the author worked in a printing plant where some workers were setting lead type without knowledge of the implications of exposure nor the skills to minimize exposure, the machine operators I know in the United States spend a great deal of time in meetings and training to keeping up with the multimillion dollar engineered systems they operate. The image of Homer Simpson, the jovial bungling cartoon character who is a safety inspector at a nuclear power plant, is ridiculous. Losses of key personnel can serious hamper noxious facility operations. In 2013, the U.S. government stopped paying its bills. At the Savannah River former nuclear weapons site, for example, some key employees left their jobs and went to work across the Savannah River on the construction of a nuclear plant (Vogtle in Georgia). Not all returned, leaving the Department of Energy's site manager with a serious problem that impacted their work schedule. Technology has taken over repetitive work in some industries. Nevertheless, noxious industries need a cadre of expert and reliable workers. If they make mistakes, serious consequences are likely to follow.

Markets and clients

Fifty years ago, it really helped to be in a growing metropolitan region to access local customers, as well as access distant clients via water or rail transportation (see Chapter 3 examples of an auto assembly plant, copper smelters, mini-steel mills, paint and pigment plants, and roofing tiles).

Today, international trade agreements have changed the location of production of copper, aluminum, and other products. But I would not assume that what we see today will be what we see in a decade. World markets are massive and growing, supported by advertising to spread consumer interest to places that not long ago had tiny markets. For example, the world has long become used to Middle Eastern nations selling oil to U.S. clients. In 2015, the U.S. became an exporter of shale gas to Canada and Mexico via pipeline and a major debate is being engaged about whether the world's former major sink for imported oil and gas should be selling liquefied natural gas to Japan, South Korea, and elsewhere.[29]

In regard to a hazardous waste market, rational thinking would lead to managing toxic waste locally in order to avoid the possibility of transportation-related accidents. However, NIABY (not in anybody's back yard) has been interpreted by some that it is allowable to ship materials for disposal in other countries and/or placed on large boats that would incinerate waste in the oceans, or even launch the material into outer space (Chapter 8). This creative, albeit I think misguided thinking, can create risk for those not living near the site. And yet, the counterpart is industrial inertia, which means that the same site is used over again or enlarged to avoid risks associated with new sites and take advantage of agglomeration economies.[30] Chapter 3 describes the Kin-Buc Landfill case where a local landfill became one of the worst in the eastern United States (see Chapter 6 for a discussion of the CLAMP policy).

Capital

Managers of privately owned noxious facilities, with some exceptions, work for large well-funded multi-national private or public organizations. Their need to borrow money from large banks or financial institutions for new large facilities is less of a problem than it is for smaller private organizations and local government.

It is comforting to think that major government facilities have access to tax-payer funds to pay the bills for managing hazardous waste and to assume control of abandoned toxic chemical and nuclear facilities. Having followed and been part of the effort to manage nuclear defense waste in the United States, I realize that government agencies with large budgets have had a difficult time keeping costs within reason (five to 10 times original estimates for complex waste man-agement programs are not uncommon), and the federal government has not been willing to allocate additional resources, with the net results that projects to manage toxic noxious wastes are delayed far beyond original plans, with $ billion plus price tags.[31,32]

Overall, public managers facing legislators and private counterparts facing impatient stockholders encounter a great deal of angst if they fail to site a facil-ity that the organization needs. Delays cost money, jeopardize existing invest-ments and future plans.

Suitable land

Manufacturers need land for processing, storage, final products, and waste man-agement. Then of course they need space for parking and transportation facili-ties. They may need land for a buffer, expansion, and for recreational facilities. Some of the land may need to be on the most stable ground, if it is in an earth-quake zone, and be elevated to avoid flood surge, chronic flooding, and a high water table.

When a site manages hazardous materials, these needs are compounded and a large buffer may be required. For example, every nuclear power plant has a mandated buffer, which varies depending upon the needs for evacuation, security, and other factors. For these reasons many noxious facilities are delib-erately located in remote locations. When noxious facilities are not located in remote facilities, the consequences can be disastrous for the company and gov-ernment agency (see Sayreville dynamite plant and Manville asbestos plant in Chapter 3).

The Bhopal chemical plant leak in 1984 is among the best known cases of picking a site that was suitable for one function but not others. The land had been zoned for a "light" industrial use and was a central location that facilitated transportation of resources and final products. But during a stressful economic period for the local company, it integrated backward to include more complex engineered systems and much more toxic chemicals on the site. A series of low probability but high consequence events occurred, including engineered system failures, that led to thousands of deaths and injuries.[33] The site was not suitable for a noxious industry.

Abandoned old industrial facilities and their adjacent land have been a major public health and environmental challenge, as well as an economic opportunity. Even when resources are available to demolish the structures, land contamination and building rubble are left, and typically contamination has transferred to nearby land and water bodies (see multiple examples in Chapter 3 and Chapter 8). Liability that follows from contaminated land can fully or at least partly undermine an organization's economic health (see Chapter 3).

Sadly, some of the worst exposures have occurred at sites that were suitable for one set of activities but not for noxious ones that had not been adjusted to reduce added risk.

Agreeable governments

Local government can send clear messages to welcome or keep out industry that go beyond zoning and building codes. Researchers have found markedly different local business tax rates in adjacent local governments.[34,35] Government can attract industry with free or discounted land, a willingness to provide local technical education programs, and offers to repay cleanup costs for contaminated sites to new developers who generate jobs and tax revenues. In some cases, government has elevated the priority of a road and bridge project in order to send a "we want you" message to an industrial or other kind of developer.

Yet, of course, local government can make it difficult, if not impossible to locate a plant, by challenging every part of an application, thereby costing the developer time and associated costs. Local government, in short, rarely has veto power when the developer owns land, but it can make the developer's efforts sheer misery during the siting process and even after a facility is sited. These favorable and unfavorable reactions are exacerbated in the case of noxious industry because local government and local constituents have strong feelings about many noxious facilities, as illustrated throughout this book.

Federal and/or state governments have the power to regulate noxious facilities, although the regulation varies markedly by nation and sometimes even within the same nation (see Chapters 4 and 7). Government often has the complication of being the owner, as well as the operator, and the regulator. National, state and local government may also be the developer and user in the case of electrical energy, water and sewer plants, and solid waste management facilities. A number of the case studies later in the book feature the government in a role that pits government against government.

International trade agreements and disagreements can be unpredictable for private enterprise. Raw material prices can change, transportation routes can be blocked or at least hindered, markets made more difficult if not infeasible to reach, and property can be confiscated.

Within a nation, the national government can support a manufacturing business by providing grants, taxes, loans, trade barriers and other benefits that help a manufacturer. Government can catalyze a manufacturer in a specific area by providing tax benefits and capital to distressed areas, such as Appalachia, and selected economically distressed cities. State and local government have done the same thing at their geographical levels. However, these agreements are created politically and can be dissolved politically. All of these agreements are potentially high benefit balanced against high uncertainty.

Infrastructure, including water and waste management

Infrastructure includes, water, waste management, communications, and electrical networks, and other essential systems. For example, beverage manufacturers require high quality underground fresh water or city water; some require large amounts of cooling water; and just about every manufacturer requires potable water for domestic on-site purposes. Some require all of these.

Waste managers have widely varying local needs. Some have discharge permits and others reuse their gray water. Some do both. Noxious industries present special challenges. In regard to cooling water, some heat the water and evaporate a great deal of water (nuclear power plants). When the facility is closed for maintenance or a failure, then water temperature can drop killing thousands of fish. Effluent discharge permits issued in the 1970s to many noxious industries with water discharge permits showed large discharges of not only heated water but also oxygen-absorbing organic and chemical products and chemical and physical hazards (see Chapter 3 for multiple examples).

As the data in Chapter 1 show, for sheer obnoxiousness, waste management facilities top of the list of essential facilities that people do not want near them. The author was not surprised when the first major case study of environmental injustice in the United States was a hazardous waste landfill, which led to a dispute about the siting of hazardous waste disposal facilities in the United States (see Chapter 4). Environmental justice issues extend to international shipments of hazardous waste, which have been controversial and generally perceived as developed nations in the northern hemisphere dumping on poor nations typically in the southern hemisphere.

Suitable local environment, including acceptable weather

Many attributes fit under this label ranging from local terrain to weather. Starting at the regional scale, some industries need or prefer a specific climate. For example, U.S. aircraft developers chose the west partly because the weather permitted all year testing, as well as the availability of large tracts of land.

At the site level, industry needs a location that allows it freedom to modify facilities. In regard to noxious facilities, placing it on steep terrain, in places where ground is not stable, where there is an unacceptable high probability of floods and storm surge, and so on is an unnecessary risk. The list of site-specific attributes for noxious facilities is long (see the Chapter 6 example of an elemental mercury storage facility).

More generally, an agreeable environment for many noxious industries starts with a location where there are no people in the area. One former colleague who worked for a company that owned nuclear power plants, tongue-in-cheek, noted, he would look for places with no colleges nearby with "busy-body" faculty members. An agreeable environment ideally should have a stable geological base, hard rock that is impervious, and dry weather, which means that raw materials, products and waste can be stored inexpensively.

Transportation

When I stood for my PhD oral exams during the 1960s, the first question was "What are the two most important factors that influence location of industrial facilities?" My response was transportation and transportation. I was not trying to be funny (I was nervous). The more options an industry has to secure its raw materials, ship its products and wastes, and recruit nearby employees the better. Water, rail, road, mass transit, nearby airports, bike and walking paths are all in play. The suppliers, the clients, and the workers all want to minimize their travel times and costs.

When I learned location theory, water transportation was the cheapest for distances of 350+ miles, rail for shipments of 100 to 350, and trucks were touted for short hauls of less than 100 miles. However, many reasons make this generalization inappropriate today, and problematic even 50 years ago. To get a better estimate of the transportation costs required knowing if the mode of transportation was only one way or was in hauling back something was an option; and was there a special arrangement that required customers to pay the same costs whether the product was shipped 100 miles or 1000 miles, as was the case for the steel industry for many years. In short, estimating transportation costs was difficult, and today figuring out the best option among larger ships, longer trains, faster speeds and other needs challenges experts.

The inference today is that not only are financial services and many other forms of business "footloose," that is, have many location choices, but that same argument has been made about manufacturing. That is, globalization has made it possible to ship many products long distances. Noxious industries do not have as much freedom of choice. In fact, I believe that many are increasingly constrained by the threat they pose and the threat that they are perceived to pose to human health, the environment, and property values.

Entrepreneurship

Entrepreneurs may be given too much credit for success and much too much blame for failure. I have observed that strong managers who do not lose sight of their key goals can make a difference when noxious facilities are involved. The processes described in Part II are in response to the challenge of finding locations that reduce uncertainty and risk for managers, their organizations, as well as local governments and their populations.

Evolution of industrial location theory and tools in the twentieth century

Economists, geographers and regional scientists with access to limited data, using physical models, slide rules and primitive calculators began developing industrial location theory. Mostly they relied on deductive reasoning grounded in micro-economic theory to explain why manufacturing facilities located where they did.

Prior to considering that evolution for manufacturing, I note that retailing location was the focus of much of the early research. Briefly, central place theory was developed by German Walter Christaller.[36] His first German-language text, examined the location of retailing and found that the hexagonal shape most efficiently packed space without leaving empty space, in other words, the hexagonal shape would emerge on a flat landscape in order to minimize distance for consumers. That work fit undifferentiated rural landscapes, but required much more to fit to suburban and urban landscapes. Much of that translation was accomplished by Brian Berry, Barry Garner and their colleagues.[37] The basic principle was that stores with the highest value products would locate in places with the largest nearby market. Central business districts would feature all of the major high values products such as jewelry stores, expensive clothes and other high value products. In areas outside of the central business district, high value product stores would concentrate at or near street intersections. For example, while a graduate student, I along with two of my fellow graduate students walked along the elevated subway lines in New York City. What we found was that the stores immediately adjacent to the exits of the elevated stations were higher value ones, such as jewelry stores and drug stores. A block or two from the stations where less foot traffic passed was where food stores, low value apparel, and others stores concentrated.

Unfortunately, as in the case of manufacturing industries, when neighborhoods lost their affluent and upper-middle-class populations, the value of their retailing declined. Hence, a great deal of store abandonment occurred in the middle of blocks and away from areas with a lot of foot traffic, exemplified by the work of Berry et al. in detailed studies of Chicago.[37] Summarizing, retailing location theory and practice is fascinating, shares some of the characteristics of noxious industry, but it is beyond the scope of this book.[37,38]

The major focus of this section of Chapter 2 is the origin of industrial location theory. The key premise is that decision-makers are rational, their organizations are rational, and both understand and have access to pertinent economic data upon which to base decisions. One of the clearest English-language articulations of this viewpoint was written in 1947 by George Renner.[39] Renner wrote the "law of location for fabricated industries." He said: "any manufacturing industry tends to locate at a point which provides optimum access to its ingredient elements"[39] (p. 181).

The ingredient elements were materials markets, power and labor.

Alfred Weber sets the frame

I focus this section around German economist Alfred Weber. A professor in Prague and later Heidelberg, he published a book in German on industrial location that guided industrial location theory for the twentieth century. Earlier in this chapter, I noted that I elicited some laughter from my PhD committee when I ranked transportation as number 1 and 2 in driving industrial location. Weber's focus on transportation was ingrained in my generation. I read portions of this book in German, clearly missing many of the subtle points in his argument, until I realized that it had been translated into English.[40]

Weber, like many of his contemporaries began with some simplifying assumptions. These were:

- A uniform climate, production techniques, and an undifferentiated population (in his words they were of the same race);
- A single product (or one product at a time). A different type of shoe would be a different product;
- Raw material locations are known;
- Markets are known;
- Labor supply and wages are known; and
- Transportation costs are related to distance and weight.

Weber divided materials into types. "Ubiquitous" materials are available almost everywhere at the same price. "Localized" materials are to be found only in a few known locations. "Pure" materials are localized materials that become part of the final product, whereas "gross" materials are not part of the final product or only a small part is part of the product. From these distinctions among pure and gross materials, the manufacturer calculates a locational weight, which is the total weight to be transported per unit of product. The locational weight is then mapped in order to calculate the best location. The analyst then calculates higher cost locations around the best site, which are called "isodopanes."

Weber then used his approach to find the location of industrial production for the case where there is one source of raw materials and one market, two sources of raw materials and one market, and so on. For readers who fancy two-dimensional geometry, Figure 2.1 presents the case where there is a single market and two weight-losing localized resources. The three locations are represented by a triangle with three points (P1, P2, and P3). Depending upon how much weight is lost in manufacturing, the location of the production (point

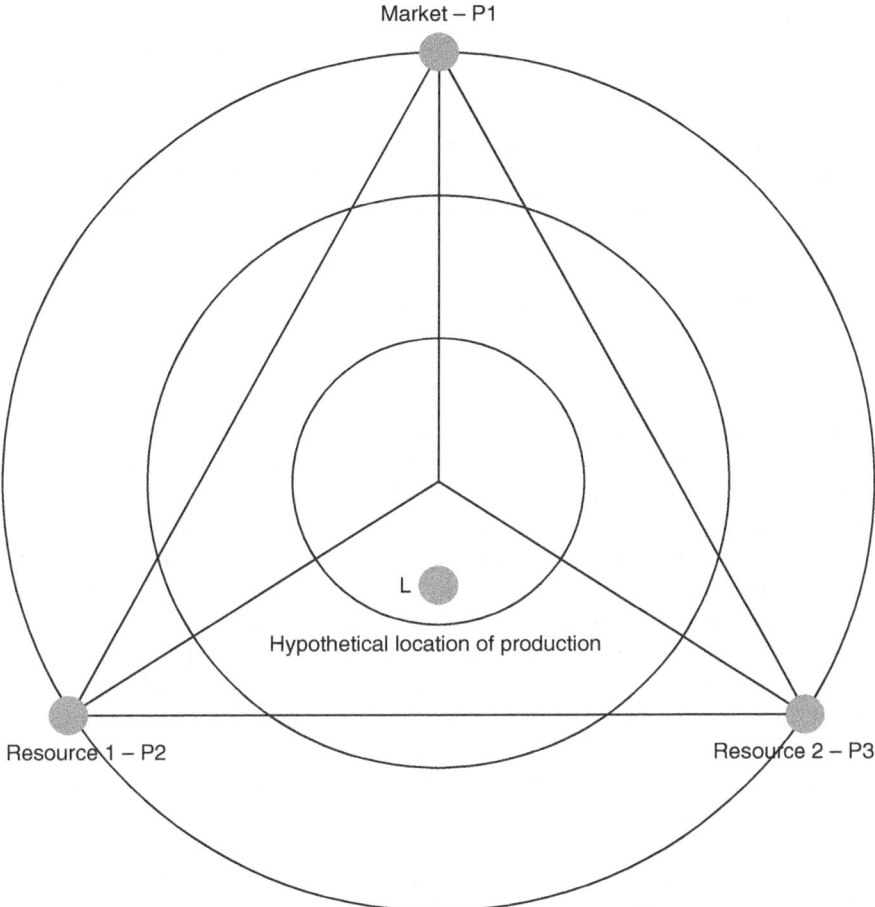

Three ellipses are isodapanes of added cost around most
efficient site. The outer one is the most disadvantageous

Figure 2.1 Basic Weber location model.

L) will be somewhere inside the triangle because the rational producer will not
want to ship raw materials that are not part of the final product – see Figure 2.1.

Weber expected the producer to pick the location that is most economical. If
the number of resources and markets increase, then the graphical ability to general-
ize the location is more difficult. However, this is where a physical analogue comes
into play, called the Varignon frame. Assume that each resource and the market
can be represented on a physical model built on a table. Each fixed point (resource,
market) represents the pull of each of these, which is accounted for by physical
weights. These weights are connected by strong string. When each weight is
allowed to pull the string, the heavier weights pull the location toward them.

The Varignon frame was a superb teaching tool. My class in graduate school built and used a Varignon frame, thereby persuading ourselves that transportation costs ruled (see Kennelly case below). Weber understood that the best location might not be possible for a location. Hence, he drew isodopanes around the optimum location to measure points of equal cost around the optimum site. In other words, these equal sub-optimal set of circles in Figure 2.1 around the optimum site measure the cost penalty of not choosing the minimum cost location. Weber's seminal book also addressed the issue of industrial agglomeration associated with building a larger manufacturing plant, building relationships with manufacturers of the same or similar products that share scale economies, and other businesses in order to reduce their cost.

Weberian case studies

Weber's book led to a string of applied industrial studies, many focusing on noxious industry locations. Lindberg[41] examined the location of the paper industry in Sweden during the years 1830–1839, 1890–1899, and 1930–1939. In the earliest period small handmills manufactured the paper. The raw material changed to groundwood mills and later to chemical pulp mills. Despite the change in source, Lindberg concludes that wood sources are so ubiquitous in Sweden that wood only played a small role in locating mills.

Smith[42] examined the location of British industry, more specifically he considered the extent to which industry location was oriented toward raw materials that shed weight before being delivered to the final customer in the final product. He observed that sugar beet plants are located in the beet-growing areas, blast furnaces in coal or ore fields, and milk processing factories among dairy farms. Each of these has major weight losses, which confirmed Weber's expectations of a resource-dominated industry. However, industry not dominated by a single resource had more options. He noted that with the spread of the British railway industry, more flexibility was created for industry, which was apparent for the iron and steel industry. The primary manufacturing stages of the industry used heavy and bulky raw materials, and these facilities remained near or literally on top of the raw materials (coal, iron). Later stage products such as nail and screw mills, textile machine shops, and motor vehicle factories were not located near raw materials.

In the early and middle of the twentieth century, steel mills were a prize industry with which to build a strong economic base. They created a lot of jobs, income and taxes (see below). Kennelly[43] studied Weber's location analysis using the Mexican steel industry. In 1950, Mexico had two large steel plants, three smaller secondary plants, and other smaller producers. These plants produced about one-half of the country's needs. Both plants were located in the north, centered on Monterrey. The key materials, based on weight and cost, were iron ore, coke, oil and scrap from markets. Labor had little influence on the locations in this era in this country. The markets were concentrated in Mexico City (about 70%) and Monterrey (24%).

Given this limited number of key resources, limited number of plants, and spatially concentrated market, the authors built a Varignon's frame. They used a board with a map of Mexico mounted on it. Holes were drilled at market and resource supply locations. Weights reflecting the weights of the supply were suspended on the strings and the strings were knotted above the board. The knot was free to move in all directions and arrived at the theoretically best site. Mathematics and geometry were used and found the same site. Kennelly also drew isodapanes around the optimum site, which allowed him to estimate costs associated with other locations. The locations determined by the physical and mathematical models were close to where the primary and secondary mills were located. Kennelly concluded that the Varignon frame and two-dimensional geometry worked for cases with limited numbers of resources and concentrated markets, but new rail connections and resource substitution could change the patterns.

Allan Pred,[44] a geographer, economist and historian, made numerous important contributions to our understanding of industrial location. Here, I summarize his analysis of the geographical concentration of high-value added production. The study is an empirical investigation of Weber's ideas about industrial agglomeration. His "cursory treatment" (his language) of 23 industries in the United States is in fact anything but cursory. As forecasted by Weber, industries such as bookbinding, typesetting, engraving and plate printing, electrotype and stereotyping, furs, and many other high value ones were found in the American Manufacturing Belt, an area of approximately 850 miles (1369 km) by 400 miles (644 km), which included the 10 largest cities in the United States when Pred was conducting this research.

Pred observed that in 1958, 47.3% of the U.S. population lived in this area. This compared with 65.3% of the nation's value added by industry. Of these high-value added industries, 17 of the 23 had at least 80% in the American Manufacturing Belt. Two high-value manufacturing industries were nonconforming to the pattern: newspapers and ice manufacturing. Each are what Weber would have said are based on ubiquitous resources, located in every large and small city. The second was also overrepresented in the South because of warm temperatures. Pred concluded that this pattern could change, and indeed it has as the American Manufacturing Belt has lost a great deal of its manufacturing and some now label it or parts of it the American rust belt because of all the old abandoned factories (see Table 2.1 above and Chapter 3 for some illustrations).

Walter Isard's[11] contributions across the field he called regional science are legendary among those of us who were fortunate enough to read his books and watch him lecture. As part of the period's tendency to focus on the iron and steel industry, Isard and Cumberland[45] examined the possible siting of an integrated iron and steel plant somewhere in New England. The authors never mentioned location theory in their study, in fact, they did what they did best, which was to analyze location data using a variety of methods. In 1950, when this study was done, major integrated iron and steel mills were located in the American

Manufacturing Belt in Bethlehem (PA), Buffalo, Cleveland, Pittsburgh, Sparrows Point (east of Baltimore MD), and Trenton (NJ). Isard and Cumberland carefully compared the transportation costs for the existing locations versus proposed ones in Fall River, Massachusetts and New London, Connecticut for iron ore, coal, and then shipment of finished products. The two New England sites enjoyed a slight advantage for shipments as far south as Boston and Providence. But this transportation cost advantage was not maintained compared to Trenton, New Jersey for markets in Connecticut. The authors concluded that an iron and steel plant in either of the two New England sites was a marginal investment.

Looking back at this study 67 years after its publication, the idea of a fully integrated iron and steel plant in New England was an investment that wisely was not made. Nearly all the steel plants, and especially the integrated mills in the Isard and Cumberland study, closed.

Changes during the last 100 years have not eliminated some basic needs. In regard to noxious industries, Rawstron[46,47] identified three location restrictions that remain just as important today as they were when he wrote them:

- Physical;
- Economic; and
- Technical.

His first restriction is that production may only be possible at selected sites because of the importance of key resources that are available at a few locations. Second, even if those key resources can be identified it may be infeasible to produce the final product economically at those sites. Yet, even if some sites can produce economically, there may be technical requirements that make those sites infeasible. Rawstron illustrated his ideas with the location of coal-fired electricity generating stations. He started with the requirement for cheap coal, rail access, and, of course, a place for waste disposal. In his example, the market for coal is not restrictive because of the extensive electrical grid to deliver the final product. However, the presence of coal clearly was the key driver. Hence in his example, the site needed to be near or even on a coal field.

Narrowing down the choices, coal plants require access to cooling water, so sites need to be near rivers in the coalfields. At that time a floodplain was a good location because it represented flatlands. It helped to have a nearby gravel pit or some other depression to dump the waste. Then, an adjacent rail line was essential.

Rawstron's analysis looks simplistic from today's perspective. However, this kind of reasoning is certainly not lost upon those who seek to locate iron and steel, chemical plants, landfills, and many others, Today, Rawstron's list of restrictions would be very long. Indeed, having worked on siting projects that eliminated locations based on sequential limitations (Chapter 7), I can say that all the sites sometimes can be eliminated!

Behavioral theory

Traditional industrial location theory was grounded in deterministic economic thinking, assuming that entrepreneurs and their firms had access to all pertinent information, behaved rationally in decision-making process and focused on profit maximization and cost minimization. During the mid-1950s economist Herbert Simon began formulating his theory of "bounded rationality" that eventually led to his Nobel Prize in economics.[48] He asserted that entrepreneurs did not have all the information, could not absorb all the information, and had time limitations, all of which precluded an optimal decision. Simon argued that in the end decision-makers need to settle for what Simon calls a "satisfying" search process not an "optimizing one." He initially applied his thinking to developing a heuristic for locating warehouses.

In regard to siting noxious industry, the idea of a satisficing solution makes a great deal of sense. Beginning with the entrepreneurs as noted in the preface, upper managers vary in their knowledge about business, engineered systems, pollution control, labor, local zoning and taxes and many other decision-making elements. Presumably, every company and government agency has access to all the requisite formal training and experience. However, based on my personal experience, they do not. Upper management may obtain knowledge that is dated, which they may not know, and further may not be willing to take the advice of people that report to them.

In the absence of accurate data on every key subject and under time pressure even the most rational and meticulous leaders create mental heuristics to guide them through complicated choices. Their mental models will include some data, but they are also influenced by their background and their personal agendas and preferences. Alan Pred[1,2] identified this kink in the idea of rational decision-making and characterized the likelihood of the optimal decision as a probability distribution.

Melvin Greenhut[49] distinguished between plant location in theory and practice. His list of key location factors includes the first four listed below that are grounded in the literature developed by Weber, others that followed Weber's approaches, and the last three personal factors that were not part of the traditional location factors:

- Cost factors (transportation, labor, processing);
- Demand factors (efforts to monopolize market segments by some firm, interdependence of firms);
- Cost-reducing factors (agglomeration or deglomeration factors that influence production);
- Revenue-increasing factors (agglomeration or deglomeration factors that affect revenue);
- Personal cost-reducing factors;
- Personal revenue-increasing factors; and
- Personal considerations for decision-making (nonpecuniary satisfaction).

New analytical tools in the mid-twentieth century

I would be disingenuous if I did not note that during the middle part of the twentieth century several methods of analysis were introduced that many of us thought had great promise in facility siting. Prior to this period, various researchers had used regression analysis, and this continued, especially as computers became more available. However, with more data being gathered and with computers, it became possible to use two economic-centered tools: input–analysis developed by Wassily Leontief and linear programming by John von Neumann and George Dantzig. The importance of their work was recognized by economic and military strategists, and all three received national and international honors. Those interested in facility siting were drawn to creating applications of their tools at the microgeographic scale.

Input–output (I–O) models can help us understand forces of agglomeration and deglomeration and the multiplier impacts of businesses and policies. Created by Wassily Leontief[50-52] who won the Nobel Prize in economics for his development of I–O, the real issue for those interested in industrial location theory is whether this tool would be useful for analysis at the local, county, state and regional scales. An input–output model is a table that shows the flow of transactions within the economy. If I want to build a mini-steel mill, I need to buy scrap iron ore and steel and have a cheap source of electric power. I will need a group to build and set up air pollution control systems to capture particles that otherwise will go into the atmosphere (see Chapter 3 for two examples). I will need access to rail, water and highways to move products.

The U.S. government collects data from businesses and periodically calculates sales and purchases among industries. An input–output model would help me understand the supply side and demand side impacted by my plan to build a mini-steel mill.

The first local scale geographic study of input–output that I am aware of was by Gerald Karaska of the Philadelphia economy.[53] Later the editor of the journal *Economic Geography*, Karaska used survey methods to separate purchases from local companies and outside ones. The essence of his findings was that these transactions are markedly different at the local scale, with variations of two to five times for seemingly the same transactions. Looking at the coefficients for paperboard containers and boxes, few links seemed similar even within the same industry. Paperboard is a big transaction, and in this case import transactions were more than double local ones. The author concluded that enormous variation exists with the same industry. Karaska's work means that we need to dig into these resource transactions to understand what role they play in siting. This carefully done study convinced me that we would need to build our own I–O models based on local data as much as possible to generate realistic forecasts for decision-makers.

Linear programming is an analytical tool that allows the analyst to examine problems that are posed as maximizing benefits or minimizing costs subject to constraints. Emilio Casetti used linear programming to examine the possible location of a steel mill in Southern Ontario.[54] The objective was to minimize the cost of shipping resources and final products along the St. Lawrence Seaway. Four sites were identified as possible locations. The cost minimization objective occurs after the following two major constraints are satisfied:

- All market demands must be satisfied; and
- Coal and ore materials must be shipped to four possible production centers to satisfy their needs.

In addition to these major requirements, the authors added constraints to make sure that ships were required to return to their starting places to support ongoing shipments. The results show that the raw material costs were more significant drivers than the market ones. However, as in the input–output model case, the use of these tools to explore this problem in the detail required for meaningful policy input awaited high speed computers. The real obstacle was computing the solution. Until high speed computers were available, optimization modeling was infeasible to use except by the U.S. military and for a few other special projects. Computers allowed me to do a PhD thesis that rerouted water during a drought to places that were running out of fresh water. In regard to noxious facilities, I was able to build a model that looked for locations for trash burning or landfilling sites. By then, we had high-speed computers, which by today's standard were crude and slow (see Chapter 9). But even today's computers cannot help when we do not have a clear understanding of the values, ethical, and political dimensions, and other data sets required to make a comprehensive assessment of siting and non-siting options (see Chapter 10).

Final thoughts

Economic-based thinking drove location analysis in general and industrial location siting for much of the twentieth century. Economists used graphs, and built physical and later mathematical models that reflected their deductive thinking about facility location. Companies and agencies did their economic and engineering studies, and announced their choices. This step was followed by filling out permits and other mandatory forms. Then they built the facilities. Even while this thinking prevailed during the "choose-announce-defend-build" era, we learned that decision-makers did not have access to all the salient economic information nor did they use all of what they did have. If profit-maximizing leaders did not make optimum or even near optimum economic-based choices for their organizations, then it is safe to assume that their choices would not lead to good results for some project neighbors, nor the surrounding jurisdictions as a whole. Nevertheless, sites were picked, and as Chapter 3 will show many started out as small facilities, grew larger as markets expanded, and were continued by sunken investments and inertia until a combination of economic, social, environmental and political forces made them appear to be pariahs.

The vastly more complex world environments of the twenty-first century have only served to compound these challenges. Joel Tarr's[55] book on the growth of the steel industry in Pittsburgh, the closing of its steel mills, and efforts to revitalize the city is a marvelous account of the growth and wealth brought by manufacturing, the abrupt ending of the industry, and efforts to bring a new economy and image to the city. The same efforts characterize efforts in Buffalo, Cleveland, Detroit, St. Louis and other American Manufacturing Belt cities that are trying to recover from building cities with manufacturing dollars and then shrinking cities when these were lost. Chapter 3 focuses on the same set of ideas in six small industrial cities.

References

1 Pred A. *Behavior and Location: Foundations for a Geographic and Dynamic Location Theory*, Part I, Lund Studies in Geography, Series B, 27, Lund, Royal University of Lund, 1967.

2 Pred A. *The Spatial Dynamics of U.S. Urban Industrial Growth, 1800–1914. Interpretive and Theoretical Essays*. Cambridge, MA, MIT Press, 1966.

3 Pred A. Industrialization, Initial Advantage, and American Metropolitan Growth. *The Geographical Review*. 55, 158–185, 1965.

4 Alexander J. The Basic–Nonbasic Concept of Urban Economic Functions, *Economic Geography*. 30, 3, 246–261, 1954. p. 261.

5 Alexandersson G. *The Industrial Structure of American Cities*. Lincoln, NE, University of Nebraska Press, 1956.

6 Blumenfeld H. The Economic Base of the Metropolis. *Journal of the American Institute of Planners*. 21, 114–132, 1955.

7 Hoyt H. The Utility of the Economic Base Method in Calculating Urban Growth. *Land Economics*. 37, 51–58, 1961.

8 Davis K. The Origin and Growth of Urbanization in the World. *American Journal of Sociology*. LX, 429–437, March 1955.

9 Morrissett I. The Economic Structure of American Cities. *Regional Science Association, Papers and Proceedings*. 4, 239–256, 1958.

10 Tiebout C. The Urban Economic Base Reconsidered. *Land Economics*, 32, 95–99, 1956.

11 Isard W. *Methods of Regional Analysis*. New York, Wiley, 1960.

12 Harris C. A Functional Classification of Cities in the United States. *The Geographical Review*. 33, 86–99, 1943.

13 Nelson H. A Service Classification of American Cities. *Economic Geography*. 31, 189–210, 1955.

14 Duncan O, Scott W, Lieberson S, Duncan B, Winsborough H. *Metropolis and Region*. Baltimore, MD, The Johns Hopkins Press, 1960.

15 Gottmann J. *Megalopolis*. New York, The 20th Century Fund, 1961.

16 Roterus V, Calef W. Notes on the Basic-Nonbasic Employment Ratio. *Economic Geography*. 31, 17–20, 1955.

17 Weber A. *The Growth of Cities in the 19th Century*. Volume 11. New York, Macmillan, 1899.

18 Vernon R. *Metropolis 1985*. Cambridge, MA, Harvard University Press, 1960.

19 Muncy D. Land for Industry – A Neglected Problem. *Harvard Business Review*. 32, 51–63, 1954.

20 Wrigley R. Organized Industrial Districts. *Journal of Land and Public Utility Economics*. 23, 180–198, 1947.

21 Murphy R. *The American City.* New York, McGraw-Hill, 1966.
22 Fuchs V. *Changes in the Location of Manufacturing in the United States since 1929.* New Haven, Yale University Press, 1962
23 Carey G, Greenberg M. Towards a Geographical Theory of Hypocritical Decision-Making, *Human Ecology,* 2, 243–257, 1974.
24 Kitagawa E, Bogue D. *Suburbanization of Manufacturing Activity within Standard Metropolitan Statistical Areas.* Oxford, Ohio, Scripps Foundation, 1955.
25 Fishman R. The American Metropolis at Century's End. *Housing Policy Debate,* 11, 1, 119–213, 2000.
26 Alonso W. Industrial Location and Regional Policy in Economic Development. Working Papers. 14, Department of City and Regional Planning, University of California, Berkeley, 1968.
27 Stafford H. An Industrial Location Decision Model, *Proceedings, Association of American Geographers,* 1, 141–145, 1969.
28 Smith D. *Industrial Location.* New York, John Wiley and Sons, 1971.
29 Palti-Guzman L. Gas Under Pressure: The United States is Ready to Export LNG, But Does the World Want It? *Foreign Affairs.* January 8, 2016.
30 Rogers A. Industrial Inertia – A Major Factor in the Location of the Steel Industry in the United States. *Geographical Review.* 42, 56–66, 1952.
31 Greenberg M, Apostolakis G, Fields T, Goldstein B, Krahn S, Matthews RB, Rispoli J, Stewart J. A Review of the Use of Risk-Informed Management in the Cleanup Program for Former Defense Nuclear Sites. Report prepared from U.S. Senate Committee on Appropriations and U.S. House of Representatives Committee on Appropriations, Washington, DC, August 2015. www.cresp.org/wordpress/wp-content/uploads/2016/05/Omnibus-Risk-Review-Report_FINAL.pdf
32 U.S. DOE 2013, Introduction to DOE Environmental Liabilities, May. Available from: http://energy.gov/sites/prod/files/2013/07/f2/Environmental-Liability-101-2013.pdf. November 2014.
33 Broughton E. The Bhopal Disaster and Its Aftermath. *Environmental Health.* 4, 2005. www.ncbi.nlm.nih.gov/pmc/articles/PMC1142333/. Accessed December 21, 2016.
34 Hoover E. *The Location of Economic Activity.* New York, McGraw Hill, 1948.
35 Orr S, Collingsworth J. *Urban and Regional Studies.* London, Allen & Unwin. 1969.
36 Christaller W. *Central Places in Southern Germany.* Translated by C W Baskin. Englewood Cliffs New Jersey, Prentice-Hall, 1966.
37 Berry B. *Geography of Market Centers and Retail Distribution.* Englewood Cliffs, NJ, Prentice-Hall. 1967.
38 Losch A. *The Economics of Location.* Translated by WH Woglom. New Haven, CT, Yale University Press.
39 Renner G. Geography of Industrial Localization. *Economic Geography.* 23, 167–189, 1947.
40 Weber A. *Theory of the Location of Industries.* Chicago, IL, University of Chicago Press (translation of German edition written in 1909), translated from German by Friedrich C.
41 Lindberg O. An Economic–Geographical Study of the Localisation of the Swedish Paper Industry. *Geografiska Annaler.* 35, v, 28–40, 1953.
42 Smith W. The Location of Industry. *I. B. G. Transactions.* 21, 1–18, 1955.
43 Kennelly R. The Location of the Mexican Steel Industry. *Revista Geografica.* 15, 41, 199–129, 1954.
44 Pred A. The Concentration of High-Value-Added Manufacturing. *Economic Geography.* 42, 2, 108–132, 1965.
45 Isard W, Cumberland J. New England as a Possible Location for an Integrated Iron and Steel Works. *Economic Geography.* 4, 245–259, 1950.
46 Rawstron E. Three Principles of Industrial Location. *Transactions and Papers, IGBG,* 25, 132–142, 1958.

47 Rawstron E. Electric Power Generation. In Edwards K, ed. Nottingham and its Region. Paper prepared for the meeting of the British Association for the Advancement of Science. Nottingham, UK, 31–314, 1966.

48 Simon H. Bounded Rationality and Organizational Learning. *Organizational Science.* 2, 1, 125–134, 1991.

49 Greenhut M. *Plant Location in Theory and Practice.* Chapel Hill, University of North Carolina Press, 1956.

50 Leontief W. Input–Output Economics. *Scientific American.* 185, 15–21, 1951.

51 Miernyk W. *The Elements of Input–Output Analysis.* New York, Random House, 1965.

52 Chenery H, Clark P. *Interindustry Economics.* New York, Wiley, 1959.

53 Karaska G. Interindustry Relations in the Philadelphia Economy. *The East Lakes Geographer.* 2, 80–96, 1966.

54 Casetti E. Optimal Location of Steel Mills Serving the Quebec and Southern Ontario Steel Market. *Canadian Geographer.* 10, 1, 27–39, 1966.

55 Tarr J. *Devastation and Renewal: An Environmental History of Pittsburgh and Its Region.* Pittsburgh, PA, University of Pittsburgh Press, 2003.

3 The growth and decline of noxious facilities in New Jersey

Introduction

Using a portion of New Jersey as an example, this chapter has three goals:

- Illustrate how traditional industrial location decision-making processes led to building manufacturing and associated transportation and waste management complexes;
- Describe how manufacturers in this region and much of North America and Europe reacted to changing worldwide economic conditions, environmental regulations, public and local government concerns that ultimately undermined their sunken investments; and
- Review how companies, local governments and populations in this region have responded to plant closings and their legacies.

In 1960, 27% of the United States' working population reported that they were employed in a manufacturing job. Populations in seven states reported between 36% and 40% of their labor force in manufacturing: Connecticut, New Hampshire, New Jersey, Michigan, Ohio, Pennsylvania, and Rhode Island. All seven of these states were in the American Manufacturing Belt described in Chapter 2 that stretched from New England, encompassed much of the Midwestern states and the northern part of the Middle Atlantic states.

I chose New Jersey to illustrate the formation of industrial clusters and their dissolution. In 1969, when I moved to New Jersey, the State had 894,000 manufacturing jobs. In 2003, after watching hundreds of factories close, the number of manufacturing jobs was 355,000, in other words, a loss of 540,000 manufacturing jobs. In 2015, the count was 238,000, a loss of another 100,000+ manufacturing jobs. Another way of looking at this transformation in New Jersey is when I was born in 1943, for every service job in New Jersey there were two manufacturing ones. In 1962, it was 1 and 1. In 1982, there were two service jobs for every production job, and in 2013, the ratio was 8 to 1, a remarkable change in six decades.[1]

The case studies in this chapter are chosen from industrial clusters within a 15-mile radius from my home. One sits in Middlesex County and the second in

Somerset. In 1960, these two counties reported 44% and 42% of their labor force, respectively, in manufacturing, in other words, they were more industrialized than the industrialized state of New Jersey as a whole. And within these two counties, I focus on six small communities that were even more industrialized than their host counties (Figure 3.1). Table 3.1 shows that industrial employment in these six jurisdictions ranged from 42% to 69% in 1960 and is now 11% to 14%. This chapter illustrates how the ground shifted under some internationally and regionally prominent industrial facilities and their host communities.

I chose these six small political jurisdictions for three reasons. One is that they exemplify the industrial location attributes that attracted producers a century ago. Second, they illustrate, sometimes painfully, the challenge of needing to reestablish a viable nonmanufacturing base, while coping with the industrial legacy. Third, I picked places and cases that I am familiar with. I know a lot more than I would know if I were not a resident of the region. Of course, I bring along my biases about the sites, but I try to be evenhanded.

Each of the case studies contains three major elements:

• Description of the political jurisdiction;
• Presentation about one or more industries and/or waste management facilities; and
• Efforts to redevelop the sites.

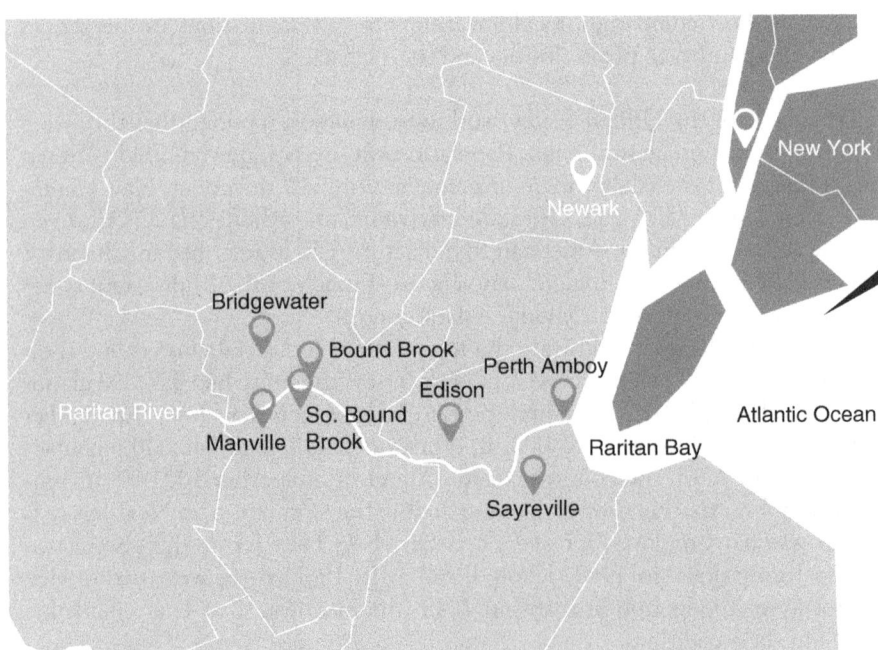

Figure 3.1 Central New Jersey locations.

Table 3.1 Summary table of industrial prominences and decline in six central New Jersey communities

Local jurisdiction and county	Substance and product	Residents of the community, % employed in manufacturing 1960 and 2012*		Population 1000s*	
				1965	2015
Edison, Middlesex	Motor vehicle assembly; Industrial landfill	42	11	56.0	102.7
Perth Amboy, Middlesex	Copper and oil refinery; Oil storage; Steel	47	14	38.4	52.7
Sayreville, Middlesex	Paint and pigments; Steel	51	11	27.6	44.9
Manville, Somerset	Asbestos	69	11	12.0	10.4
Bound Brook, Somerset	Pharmaceuticals	53	11	10.3	10.5
South Bound Brook, Somerset	Roofing products	57	14	4.1	4.6

* Sources for the data in these columns in U.S. census data files.

After presenting the case studies, the chapter concludes with a list of the factors that logically guided managers to choose these locations and factors that forced the local landscape away from manufacturing. As context, everyone of these six would be a "manufacturing" community by Chauncey Harris's[2] definition of a manufacturing city described in Chapter 2. A caveat is that the managers of these facilities have passed away, and I was only able to find limited company or personal records of why they made the location choices that they did. Hence, I am relying on both inductive and deductive reasoning, and some notes that I took when I visited or passed by those sites.

Middlesex county cluster: Edison, NJ, the Ford Motor Company Assembly Plant and Kin-Buc Landfill

The political jurisdiction

Edison, Perth Amboy and Sayreville are neighboring towns in Middlesex County. Each has a long industrial history (Figure 3.1 and Figure 3.2).

Edison is the most populous, diverse and affluent jurisdiction among the six described in this chapter. With a population of over 100,000 Edison is known as the place where Thomas Edison's creative genius led to over 1000 patents including the incandescent light bulb, the phonograph, a motion-picture

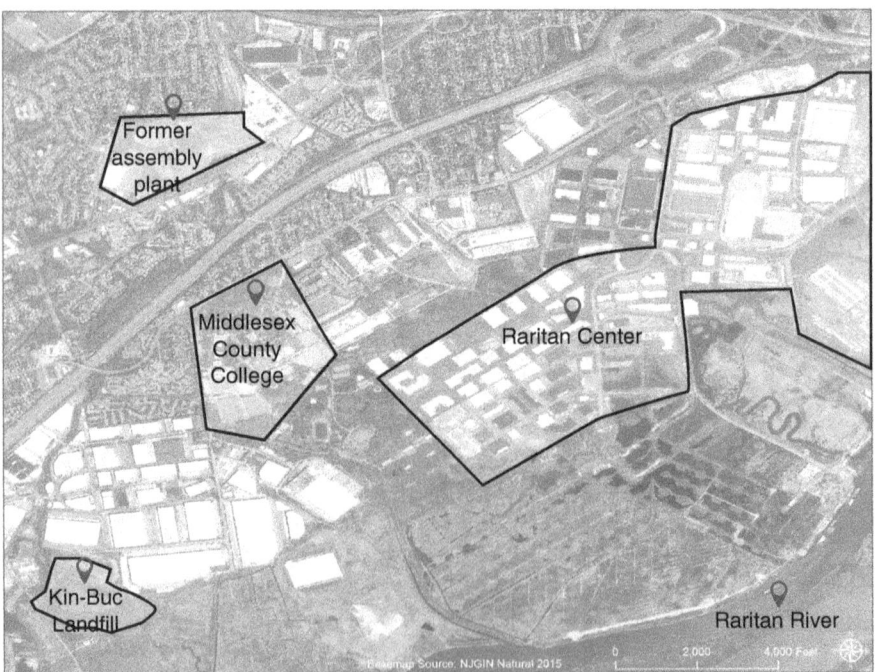

Figure 3.2 Edison sites.

camera, the alkaline storage battery, and the first electric light, among others. Decades later, tourists visit the Edison's Menlo Park laboratories where illustrations of his work are exhibited.

In 2018, Edison is one of the most demographically diverse jurisdictions. Large Asian Indian and Chinese American populations live in Edison and comprise about 45% of the population. Caucasians are about 45% of the population. The town has a notably positive national reputation. In 2006, a Morgan Quitno Survey of safe cities Edison ranked 23 out of 371.[3] In the same year, a *Money Magazine* survey ranked Edison as the second most livable small city in New Jersey and ranked 28th in the United States.[4] A 2008 *Money Magazine* survey ranked it 35 out of top 100 places to live in the United States.[5] A 2009 analysis by *U.S. News & World Report* ranked it one of the ten best places to grow up in the United States.[6]

Edison, however, is not without a challenging industrial and waste management history. I suggest that readers have Figure 3.2 in front of them as they read this section. It shows that these sites are close to one another. Two of these, the Ford Motor Company Assembly Plant and the Kin-Buc Landfill are featured here.

Before moving to these two, I summarize a very special location decision that fits classical transportation and land availability criteria, but it was not a manufacturing or waste management facility per se when founded. The Raritan Arsenal was a 3200-acre site (1295 hectares) opened in 1918 to store and transport ordnance to World War I European battlefields. The location was obvious. An enormous amount of space was available, no population settlement was nearby, ordnance could be shipped by rail and road and then stored and sent then to Europe. It would have been difficult to find a more convenient location.

After World War I ended, the site continued to function, until 1964. There were explosions on the site, but the long-term negative legacy that continues is that some ordnance was buried on the site and found during excavation for redevelopment. Some of these are still to be remediated by the U.S. Army Corps of Engineers. Having been on major DOD and DOE sites, I have not been especially worried about hazards when I have visited the Raritan Arsenal site and see areas where access is not permitted. Of course, I must confess that I was distressed when a colleague who teaches at Middlesex County College (the college occupies about 200 acres of the site) called me in April 2015 with a report that they had found a 500-pound bomb while excavating. I asked him if there was a detonator, and he called me back to say that there was not. The bomb was removed and detonated at an army base. Earlier, I had learned that United Parcel had found World War I chemical weapons (mustard gas) in 1996, and that hundreds of other rounds and casings were found buried, leading the Army to survey and clean up the entire site. But there may be some additional surprises to be found.

Despite the ordnance issue, arguably the massive site has been the most productive reuse, certainly in this book. Raritan Center is the largest industrial park east of the Mississippi River. The site contains Middlesex County College, a county college with 12,700 students. The U.S. EPA has laboratories in Edison,

Federal Express and United Parcel Service, and a number of hotels and exhibit centers that are widely used for activities such as the governor's post-election ball and boat shows, among others, and there are massive warehouses (Figure 3.2)

Overall, the Raritan Arsenal site was adaptable for retailing and wholesaling purposes, and over 13,000 jobs have been created there. Yet, at the time it was selected as a military facility it posed an enormous risk to the area. The industrial facilities that are reviewed in the remainder of this chapter brought jobs, income, economic stability and pride to their communities. Yet, in three generations the arsenal site became the favored economic son and the job-creating factories became pariahs. Readers of this book must assume that what is now desirable and acceptable may not be so during their lifetime. To assume otherwise is wishful thinking, as this chapter illustrates with painful examples.

The Ford Motor Company assembly plant

The automobile industry, perhaps more than any other, epitomizes the growth of capitalism in the United States. Cities like Detroit, Dearborn, Kansas City, and others obtained raw materials and built vehicles. These were shipped, along with electronic components, tires, and other elements and assembled, painted, in various locations across the United States. It made economic sense and provided hundreds of thousands of high paying, unionized jobs – until competition arose.

New Jersey had been one of the states where inventors attempted to build motor vehicles. When Michigan won the development competition New Jersey became one of the automobile assembly centers for Ford and General Motors. New Jersey was clearly not chosen by accident. It had clear advantages in transportation, specifically its deep water access, railroads, and large available parcels of land. The entire massive Northeast United States market is within a 350-mile radius. Water, sewer service, and electric power were available. Waste management was not a problem, at least not then.

In 1948, Ford built a plant in Edison (Figure 3.2) and another near the New Jersey–New York border in Mahwah in 1955. Both sites provided high paying jobs and ample tax revenues. However, the Mahwah plant, which closed in 1980, left serious water pollution issues. Briefly, the company acknowledged landfilling paint sludge and other wastes in a jurisdiction adjacent to Mahwah. The seriousness of the problem is indicated by the designation of the site in Upper Ringwood as a Superfund site (National Priority List). Ford has spent millions of dollars on the cleanup. The company was sued, and this was no ordinary lawsuit because Robert F. Kennedy, Junior, the firm of the late Johnnie Cochran, and the Aaron Brockovich group were all involved in a lawsuit. The plaintiffs argued that Ford and its consultants dumped hazardous waste that among other things caused cancer, diabetes, and other public health impacts over the 500 acres (about 202 hectares) dumpsite. There were countersuits and a variety of other legal actions that are ongoing and seemingly endless following the Ford Mahwah assembly plant. Judged by testimony in the various lawsuits, Ford had really not seriously considered the environmental and public health implications of locating in the affluent Mahwah area, nor siting a landfill in an area where a large proportion of the population depend on groundwater for their drinking supply. I cannot imagine that any other major auto assembly plant would make this mistake today, but I have been surprised before.[7-11]

The Ford Assembly Plant in Edison, in comparison, was considered to be a good neighbor, that is, until it chose to close its 100+ acre (40.5 hectares) facility located along Route 1 (the road that runs to the east of the site in Figure 3.2. Ford manufactured about seven million vehicles starting with the Lincoln and then including the Escort, Falcon, Mustang, Pinto and then the Ford Ranger pick-up truck.

Even after announcing that the plant would be closed, workers I spoke with highly praised the company as providing good jobs including high wages, benefits and retirement packages. Indeed, many of the ones I spoke with understood that Ford needed to rethink its assembly business; they never expected it to be at the Edison assembly plant that would close.

Suffering from competition by Japanese and European automakers in January 2002, Ford announced that it was cutting stock dividends, reducing prices for its vehicles, writing down assets and cutting 35,000 jobs worldwide including in Edison. The plant closed. Some of the 1500+ on-site workers were able to transfer to other Ford plants, others retired and received a lump payment and their pensions, and still others were discharged with three years of wages.

Ford did leave the site with a controversy. When the facility was demolished, Ford hired companies to dispose of the concrete from structures. The concrete was to be used in future developments. The concrete mix, however, that was placed at seven residential sites violated residential standards for polychlorinated biphenyls (PCBs). Multiple lawsuits later, Ford was ordered by the New Jersey Department of Environmental Protection to remove the contamination from the sites.[11]

Redevelopment of the Ford site

After attempting to change Ford's collective mind without success, Edison's plan has been to turn the 100+ acre site into a town square with a movie theater restaurants, shopping, and community center. Hudgins[12] sees the Ford plant as an opportunity to turn "idleness into opportunity." So far, a 140,000 ft² discount big-box facility has been built, a golf driving range and three restaurants opened. Given the location, despite the high cost of doing business in the New Jersey–New York region, it is difficult to conceive of this site not being fully developed within the next five years. I do not expect it to have the kind of high wages, high benefit, and long career jobs that were there for decades, but those are hard to find anywhere.

The Kin-Buc Landfill

The Kin-Buc Landfill lies directly south of the former Ford assembly plant. In 1981, as part of a class project, my students were given the assignment of choosing locations for landfills and/or incinerators. Their first job was to make a list of locations that absolutely had to be avoided. After city neighborhoods and nearby schools, hospitals and airports, they listed underground water supply and coastal wetlands.[13] Their next assignment was to map the locations of existing landfills and incinerators in New Jersey. To their dismay they found that landfills disproportionately were located in areas that are used for ground water and wetlands. This gave me the opportunity to discuss the "out of sight out of mind" principle about waste management facilities, which I have found to be the case all along the Eastern Seaboard of the United States and so many other places that I have visited across the globe. From the post-1970 perspective how ludicrous it seems to locate waste disposal facilities in some of the ecologically richest lands on a planet. In reality, however, I must confess that if I was a waste manager with a limited budget, I would have made the same decision. From the 1930s, 1940s, and 1950s perspective, if you can pick up thousands of tons of garbage, truck them a short distance and dump them in what people considered a "wasteland," you were doing your job. Even better, because they were wastelands, there was no reason to put liners to capture the waste, venting structures to prevent methane from building up and moving laterally, or investing to prevent the consequences of landfill structural failures

With 20/20 hindsight, the Eastern Seaboard wetlands should look a lot different than they do today and be more productive than they are. But this is one option that no longer exists, although there has been some localized restoration. What is done is done, and the management question is what to do with the legacy. Several of my former students are engaged in rebuilding parts of devastated wetlands and they remained absolutely dumbfounded by how my generation and previous ones allowed these ecological treasures to be destroyed. The area along the Raritan River adjacent to the Kin-Buc site is exemplary of this mistake (Figure 3.2).

Kin-Buc was a privately owned landfill and Edison a municipal one across the river from Kin-Buc. I am focusing here on the Kin-Buc site because it received a considerable amount of industrial waste not only from New Jersey but from various locations along the East Coast. The EPA estimates that 90 million gallons (U.S., 340,000 m³) of all types of hazardous industrial waste and municipal garbage were disposed of at Kin-Buc. Surrounding creeks, tributaries of the Raritan River, were heavily contaminated by a variety of hazardous materials.[14,15]

The Kin-Buc site opened in 1947 and primarily received municipal waste and some industrial waste. In 1971 it was approved for liquid and solid wastes, and by 1979 lawsuits were filed against the owners. In 1980, EPA took the lead in removing waste from the site and bringing it to safer locations and cutting off the paths that were allowing waste to flow into rivers from the site. I have collected reports that document the substantial degradation of the site and the nearby water bodies. The site has gradually been brought under control, at least on the surface. I could demonstrate this point with tables. However, my students and colleagues suggested that my personal accounts of Kin-Buc taken 42 years apart are a much better way of providing a feeling for what the site is really like (Text box 3.1)

Text box 3.1 Kin-Buc Landfill, August 1974

Kevin [my neighbor at that time] and I loaded the remnants of my termite-infested garage on to his truck and brought it to the Kin-Buc Landfill. We drove along Woodbridge Avenue for about two miles driving on a bridge over Route 1 where I could see trucks exiting and traveling along residential streets, next to the firehouse and a school to reach Meadow Road. How weird "Meadow" Road and "garbage trucks!"

The traffic light at Meadow Road stops all of us, and when it changes to green we turn right onto Meadow Road. For about a quarter of a mile we drove along an attractive looking residential street. This is only a two lane road. Some residents parked their cars on the street. They must be crazy. Trucks containing what I assume is liquid waste are clearly finding it difficult to avoid colliding with the parked cars and truck heading toward them. I said to Kevin: "Are we going to get there in one piece?"

We drive down the hill and back up a hill crossing over the New Jersey Turnpike and once again descend toward a large landfill. Once we got there they check our load and tell us that we should go up the hill and turn right. We waited with trucks in front and in back of us. Then they directed us to drive to the right side of the face. This working face is covered with garbage. White is the dominant color, but the major image is of gulls flying from location to location, men directing traffic and of the bulldozer operator moving around the face and pushing the waste piles and digging holes in it. It's really chaos up here. Kevin and I dispose of the waste as fast as we can, but we are stopped as a large industrial truck comes up to the site. It dumps a half a dozen drums out of the back, one of which splits open and light blue liquid spills all over the waste. This doesn't seem to bother anyone, but it really bothers me, and my anxiety increases as the bulldozer driver comes over and pushes all the drums into the trash and then backs up and is heading toward us to cover our contribution with more waste. We drive away from the face down the hill. I'm thinking to myself that I would not want to have any of these jobs. I hope these guys are well paid.

Subsequent events reinforce my perceptions and fears about the site. The landfill accepted industrial, municipal, and institutional waste of all kinds, that is liquid, as well as solid waste and EPA reports that the waste came six days a week from two to 300 miles away. The landfill was approximately 80 to 90 feet high and pits were dug several feet into it with a radius of about 35 feet. Liquid waste (as I had observed) was dumped directly into these pits.

The site suffered multiple fires and a bulldozer operator was killed by one of them about shortly after Kevin and I were there. Various spills occurred, and numerous complaints were forthcoming. EPA's inspections showed problems much worse than I was able to see when I was on top of the landfill. Leachate percolated through the landfill into the rivers, liquid waste poured over the side of the landfill into the rivers, and EPA reports characterized the waste as forming a black oily material. EPA's reports also provide photos of vapors emerging from the waste piles, which they assert resulted from disposed liquid chemical waste reacting with each other. In other places, they reported strong ammonia odors and other vapors that were extremely irritating to the eyes and nose. Indeed EPA staff used air packs on subsequent visits. The black material was observed entering the rivers including the Raritan, which was reported as being stained black for a considerable distance downstream. Readers who seek more information should read EPA's many reports.

On September 3, 2016, I visited Kin-Buc again. Here is the record of my notes in a second text box.

Text box 3.2 Kin Buc observations, September 2016

I walked to Kin Buc from my home, turning right on to Meadow Road. The residential area looks fine, and clearly some owners have invested in new fencing, porches and coats of paint. During the 10-minute walk through the residential area, I didn't see any large industrial waste trucks, I head back up the hill and crossed over the New Jersey Turnpike and then back down the hill toward Kin Buc. When I reached the Kin Buc entrance, a sign tells me that it is a closed landfill. No one is permitted entrance without permission. Continuing down the hill, the Edison Boat Basin in front of me, two men are putting their boat into the water. I cross over the small Mill Brook Bridge and a path that takes me along the Raritan River. I don't see anything oozing into the river. The path is roughly 2000 feet long and surrounded by salt water plants and seemingly benign. I can't see the engineered structures, but they capture the leachate that continues to flow through the landfill. I'm not worried about being hit by a truck, inhaling and being overcome by a chemical agent, or being killed by a fire.

Redevelopment of the Kin-Buc site

I see no possibility of this area changing. The Kin-Buc site should not be disturbed. The engineered structures must be protected. Back up the hill, between the river and the New Jersey Turnpike are a chemical plant, which itself is undergoing remediation; a diesel equipment plant; two recycling plants; and a massive site where they are producing mulch. The little park along the river is a jewel for people willing to walk along it. But the telling signal for me is that someone has left paint cans immediately adjacent to the entrance of the locked Kin-Buc Landfill. They see it for what is was, a garbage site.

If we could start over, I would reserve the Kin-Buc site as an ecological system. But at this point my views are irrelevant. It is what it has become and won't change for the foreseeable future. The location-decision process was short-sighted, presumably focused on short-term profit maximization, and con-sistent with thinking at that time. This place is a painful reminder of how to back bad noxious industry location decisions.

Middlesex county cluster: Perth Amboy, refineries and a steel plant

The political jurisdiction

Perth Amboy sits on the edge of the Raritan River and the New York Bight with direct access to the Atlantic Ocean (Figure 3.1). It was settled by Scottish colonists and served as provincial capital of New Jersey until 1776. Walking through city hall, Perth Amboy's historic past is apparent in drawings and paintings of former leaders. The most obvious attraction for European settlers was that it could host large ships of that era and lower Manhattan in New York City was a 30 mile boat ride away. Perth Amboy attracted new migrants. The six square mile (15.5 km²) city's population was 42,000 in 1920, dropped to 38,000 for three decades, and has increased to about 53,000 making it one of the most densely populated cities in New Jersey. About 80% of the population self-identifies at Latino, among the highest in the state. With many young poor people, Perth Amboy faces the kinds of social, education and political challenges of more populous cities.

In regard to manufacturing, Perth Amboy epitomizes a small city that attracted manufacturing and transportation facilities. By the mid-nineteenth century, Perth Amboy had kilns for manufacturing Terra Cotta, two copper refineries, and Perth Amboy became the coal shipping port for the Lehigh Valley Railroad. It was a place where resources could be marshaled and products shipped all along the Eastern Seaboard of the United States. Perth Amboy could have been a text book case of what traditional economic location theory pro-duced in regard to noxious facility sites.

Chevron's oil refinery and oil storage facility

The first oil refinery in Perth Amboy was built in 1920. Chevron purchased it in 1945. The site produced gasoline and heating oil until 1983, when the company decided that it could no longer compete in these markets. The plant downsized and switched to asphalt, which was not a strong earner and experienced some environmental spills. The asphalt plant was closed leaving about 25 employees on the large site (Figure 3.3).

The refinery site was not finished as an industrial complex. In 2012, the Buckeye Partners purchased the refinery for $260 million and invested another $200 million to modernize the site. The site has four million barrels of oil product storage. Buckeye operates 6000 miles of pipeline and needs storage capacity on the east coast to supplement its storage facility in Linden, located north of Perth Amboy's. Buckeye's plan is to ship crude from the Midwest and from its hub in the Bahamas.[16,17]

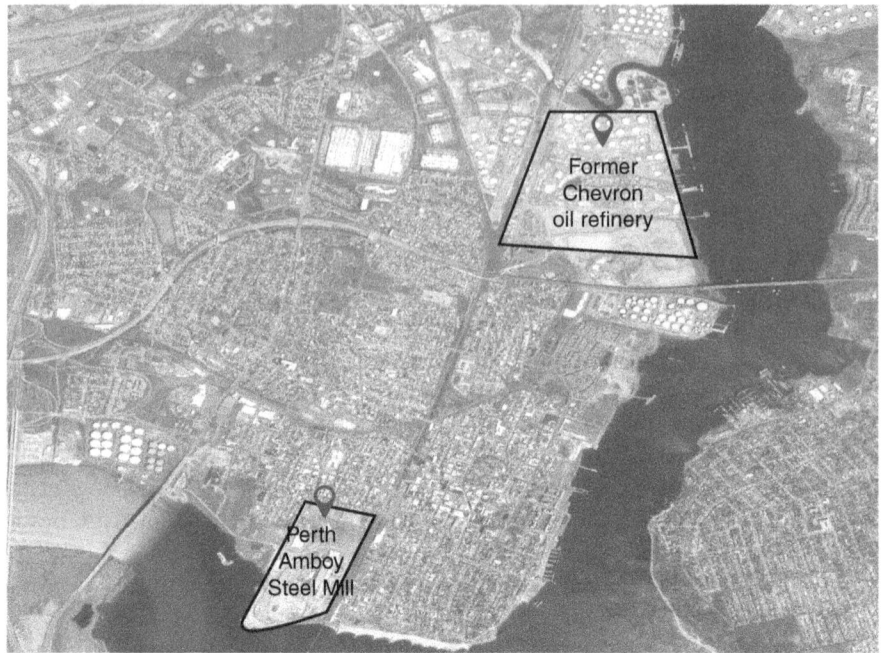

Figure 3.3 Perth Amboy sites.

Redevelopment of the Chevron site

Mayor Wilda Diaz of For Perth Amboy characterized the Buckeye plan as a new growth engine for the community. Mayor Diaz was reflecting on the reality that the Chevron refinery, which occupies about 10% of the city's land had not only become a visual eyesore but also was not producing any local jobs or tax revenues. The author had served as the volunteer external member of Perth Amboy's brownfield revitalization committee. Several members of that Committee argued that Chevron was only keeping a token workforce on the site in order to avoid cleaning it up (mothballing the site). Part of the antagonism led to a law suit against the company, and Perth Amboy was granted uncontaminated land at the edge of the refinery that was turned into commercial taxpaying land uses.

Buckeye's plan is to tie the Perth Amboy and Linden storage facilities about six miles away with a 16-inch pipeline. At a time when oil and gas are being shipped across the United States in pipelines and by train, the new capacity at Chevron will not provide as many jobs as the old refining complex, but the Buckeye pipeline project certainly is better than living with a chronic underutilized eyesore.

This reuse is one noxious facility replacing another, but there seems to be no opposition and there is some official support for local officials. I acknowledge that I have no public survey data to support my observation, only official reports and media coverage.

From copper to steel mini-mills

In 1895, the Guggenheim family built a copper refinery in Perth Amboy, and shortly after merged with American Smelting and Refining (ASARCO). This facility operated until 1975. The Raritan Copper Works, which became part of Anaconda Copper, also operated a refinery in the city.

John Gross,[18] who spent decades in the copper industry, explained the rapid expansion and decline of the U.S. copper refining industry. He says that the massive local market in the Northeast, the availability of scrap copper in the area, and the ability to ship the product led to major facilities in New Jersey, Connecticut, Rhode Island, Massachusetts, and New York. This case mirrors traditional location theory expectations. But he notes that the world market grew, changed, and trade agreements led to China and other nations emerging as the major copper producers.

The copper refinery on Elm Street in Perth Amboy closed in the 1970s, and was purchased by Gerdau Ameristeel, the second-leading mini-steel producer in the world. President Jimmy Carter attended the opening of the Perth Amboy mini-mill. The company received a major tax reduction, and low electricity rates, again the type of arrangement expected by traditional location theorists. While the Perth Amboy plant generated noise and air emission complaints and violations, it seemed like a good industrial reuse of the old copper mill site. In fact, Gerdau Ameristeel purchased abandoned property in Sayreville located near the National Lead plant in Sayreville (see below for a discussion of Sayreville).

In 1979, the mini-mill looked like the savior of the moribund U.S. steel industry. Massive integrated steel mills could not compete with these smaller mills for standardized steel products. The mini-mill uses scrap steel products, and it could compete for markets 300–500 miles from the factories. Greenhouse[19] touted the mini-mill as "steel's bright star" and in his story used the Perth Amboy mini-mill as his illustration. In 1970, less than 10% of U.S. steel was produced by mini-mills, and by the year 2000 it rose to more than 50%.

Logic suggested that the mini-mills could serve and dominate the NY–NJ–Philadelphia region of 26 million people. The Perth Amboy and Sayreville sites are centrally located in this region, each has 100+ acres located on water and has railroad spurs, cheap energy, and tax breaks.

But the mini-mill market weakened when construction projects did not grow as rapidly as expected and when the auto industry and rail cars used less steel in vehicles. The Perth Amboy mill shop closed in 2006, followed by its rolling mill in 2009. The fabricating shop remained open for a while longer.[20–22]

Meanwhile, while the Perth Amboy mini-mill was closed, the Sayreville one was refurbished.[22] It was providing a good deal of the steel for the rebuilt Tappen Zee Bridge located 63 miles away in the Hudson River. The Sayreville mill seemed to have a good market share for long steel rods for use in bridges, roads, buildings, and other concrete structures. It also has a narrow product target, can deliver products just in time, and was the first steel mill in the U.S. to receive ISO 50001 certification because of it energy efficiency. The Sayreville mill may have a future, but the Perth Amboy one does not.

Redevelopment of the steel mill site

The copper and steel site in Perth Amboy is too well located to sit idle. One option was to slightly remediate the site back to an acceptable industrial standard and reuse it for manufacturing or warehousing. The warehousing use is in great demand in this region and absorbing a great deal of formerly contaminated industrial brownfield sites. The alternative is to turn the site into housing or retailing. Perth Amboy has been pursuing the last option along its waterfront, with new restaurants, an art gallery, and boardwalk, all near its arena. On July 29, 2016, Gerdau sold the Perth Amboy steel mill site to a real estate developer, which is consistent with Perth Amboy's changing image of its future away from noxious industry manufacturing and toward its view to the east which includes the Atlantic Ocean and lower Manhattan.

Middlesex county cluster: Sayreville, NJ, titanium dioxide production

The political jurisdiction

Sayreville is an 18.7 square mile (48.4 km) borough with almost three square miles underwater. Like it neighbors Perth Amboy and Edison, Sayreville has immediate access to the Atlantic Ocean, which has proven to be both a blessing and vulnerability (Figures 3.1 and 3.4).

Sayreville was an outstanding location for noxious manufacturing. The borough is named after James Sayre, who co-founded the Sayre and Fisher Brick company during the 1850s. Given local clays deposits and location, it became one of the largest brick-manufacturing facilities in the world and the facilities stayed open until 1969, that is, 120 years.

Before the turn of the twentieth century, DuPont built two plants in Sayreville to produce gunpowder. These were important for the U.S. arsenal, and rapidly expanded during World War I. But in 1918, one part of a plant exploded killing over 100 people. The town is vulnerable to tropical storms. Hurricane Sandy inundated the area, and subsequently over 250 homes in the floodplain have been purchased for open space as of the writing of this chapter in 2016.

The NL paint and pigment plant.

National Lead (NL) is known for its lead production (Figure 3.4). It opened its first plant in the area in 1934, which produced paint and pigments. The almost

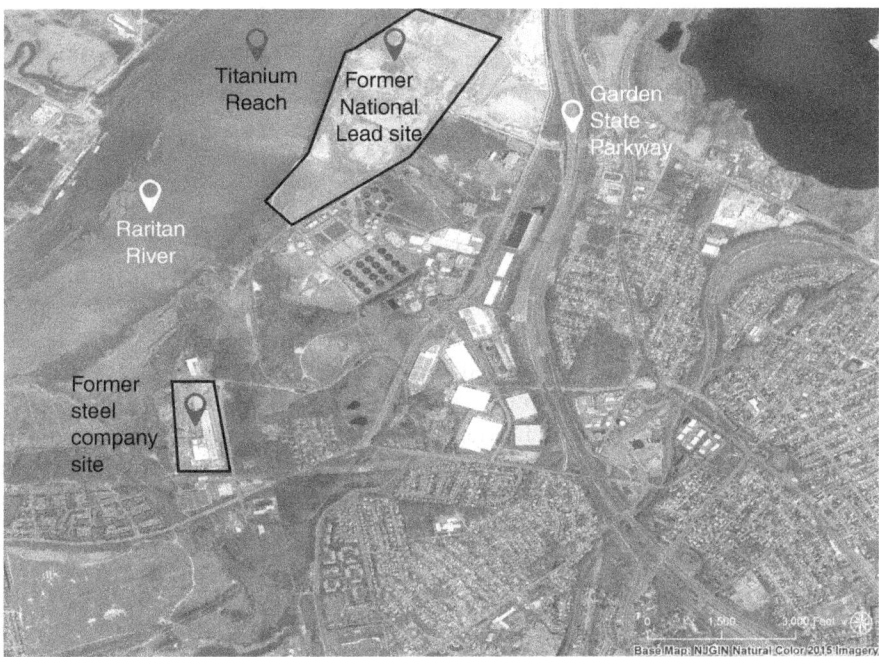

Figure 3.4 Sayreville sites.

600+ acre site produced titanium white pigments. In 1968, when I was first learning about this facility, over 1700 people were employed at the NL site in Sayreville.[23] The titanium came from Ilmenite a mixture of iron and titanium ore found in the Adirondack Mountains in upstate New York about 300 miles away. It was inexpensive to transport the mineral to Sayreville. NL built what was to be the largest titanium dioxide plant in the world in Sayreville.

The major on-site activity was making sulfuric acid, which was a challenge. This was done by bringing in sulfur, burning it to make sulfur dioxide, then adding water to the mix so it became sulfuric acid. This was boiled to upgrade the quality to about 99% sulfuric acid (really nasty material). The acid was used to treat the titanium ore, producing titanium dioxide. The TiO_2 was shipped to paint manufacturers. A key point here is that this pigment replaced lead oxide white paint, in other words, this was a positive step to protect human health. Additional key raw materials were brackish water from the Raritan River estuary and the ability to draw fresh water from on-site wells. The basics of Weberian industrial site location were on display in Sayreville.

The titanium dioxide part of the facility stopped operating when it became cheaper to ship the acid to Mexico and the sulfuric acid facility closed when it became cheaper to buy it elsewhere. The main environmental concerns for years were the so-called "acid plant" residuals. When it rained, the air emissions would come back to earth, which meant that nearby buildings and vehicles would be chronically doused with mild form of sulfuric acid. A neighbor of mine told me that the sulfuric acid mist and rain ruined the coating on the top of his car. NL responded with an engineering solution and, in fact, and in 1982 NL received the National Environmental Industry Award for Excellence in Air Pollution Control.[24,25]

The liquid wastes were stored in a clay lined basin and visible from the Garden State Parkway. The waste, which was mostly weak sulfuric acid and ferrous sulfate, contained some unrecovered titanium dioxide. A friend of the author told him that the acid plant more than doubled in size during the 1950s and the company had to build a large barge to accommodate the added waste. A few times each week, the waste, which was characterized as like "pickle liquor" from steel mills, would be pumped into a barge, taken east and south and dumped into the New York Bight.

The site for a time had a chloride plant, which produced less distressing air emissions, but used far more complex technology, and the plant did not perform acceptably because the plant operated in batches and could not easily solve process problems. Indeed, maintenance at both the sulfur and chloride plants were a problem.

A labor strike occurred in 1976, which reduced output by more than half. When the strike ended, a great deal of damage had been done by the weather and lack of maintenance. In May 1982, after much debate, the U.S. EPA ruled that NL need not relocate its dumping site to 106 miles off the coast instead of 15 miles.[25,26] At that time the plant still employed 670 people and relocating the dumping site was predicted to cost $61 million over four years. The problem went away when the plant closed in 1982.

Overall, this plant had outstanding access to raw materials, a massive surrounding market, but other manufacturers outside the area were able to compete with this site, labor costs drove up the prices of doing business at the site, and environmental regulations descended on the plant making it uneconomical to operate any longer.

Redevelopment of the NL site

Thirty-five years after NL closed its facilities controversy remains. Sayreville developed an ambitious plan to replace the NL site with a shopping mall and housing. In 2008, NL's land was sold to a developer, and NL was freed from cleanup of the on-site contamination but not the sediments in the river. A developer proposed to turn the NL site into a hotel, three million square feet of retailing, and 2000 housing units, as well as a waterfront promenade and marina. This proposal is not irrational. Millions of people drive by the location on their way to the New Jersey Atlantic Ocean shore.[27,28]

However, in this area, nothing occurs without controversy. In 2009, the EPA called for three waterfront areas near the NL site to be closed because of high lead levels. In 2010, the Raritan Baykeeper and Edison Wetlands, Inc. sued NL because of contamination left by the NL facilities in the sediments. After a study by the State of New Jersey's Department of Environmental Protection (DEP), the judge ruled in favor of NL. Two important conclusions were that the contamination had migrated from two nearby other facilities and that cleanup of these sediments needs to be part of a larger regional approach, according to the United States District Court.[27,28]

The eventual reuse of this site remains uncertain because there is a chance of debate over cleanup requirements, but the town is clearly moving down a path toward commercial and residential use taking advantage of transportation access and local views of the Atlantic Ocean and New York City. In early 2018, I learned that the ambitious plan has been scaled back to a housing development. The market at this time merits a more limited project.

Somerset county cluster: Bound Brook and Finderne, American Cyanamid chemical complex

The jurisdiction

Bound Brook, Finderne, South Bound Brook and Manville are in Somerset County about 20 miles from the three Middlesex County towns (Figures 3.1 and 3.5).

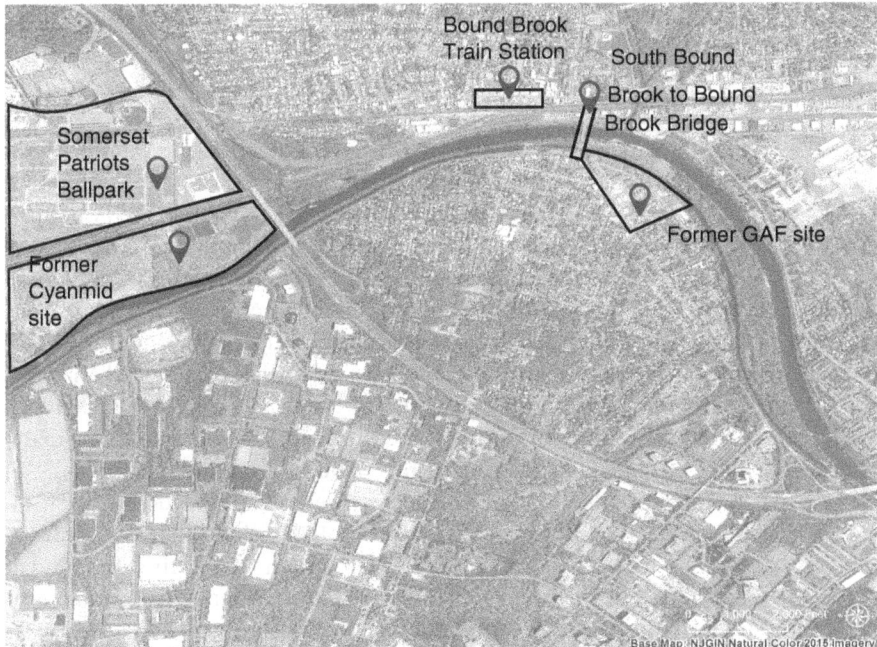

Figure 3.5 Bound Brook area sites.

Bound Brook is a 1.7 square mile borough of 10,500 people located adjacent to the Raritan River. About half of the population is Latino, and much of it is relatively poor. Finderne is unincorporated but is a census area adjacent to Bound Brook. Finderne is actually located within the much larger Bridgewater Township. However, it is much more like Bound Book than affluent Bridgewater. Main Street (Route 28) runs through Bound Brook into Finderne. In 2015, Finderne's population was about 5500 and increasingly Latino.

In contrast to these two low-lying riverside communities Bridgewater has about 40,000 residents spread out over 30 square miles. While Bound Brook and Finderne's population have been stable, Bridgewater's rapidly increased and has become the home for middle and upper middle income suburban residents, many of whom work for information industries and pharmaceutical companies in research.

Bound Brook and Finderne shared the historical benefits of industrialization and a great deal of the flood risk and environmental costs of the former American Cyanamid Bound Brook site, which is a focus of this section of the chapter. For context, Main Street Bound Brook was severely flooded in 1938 (twice), 1955, 1971, 1973, 1996, 1999, 2007, 2010, and 2011. In 1999, Hurricane Floyd sent 14 feet of water into the home of the security guard in the author's building. He had to be evacuated by boat. Fires broke out and New York City sent fire boats to help evacuate people. Between 1999 and 2015, the U.S. Army Corps of Engineers made major investments to reduce the flooding risk to a 1 in 150 year flood. Nevertheless, the chances of local flooding cannot be removed from residents' minds.

The American Cyanamid chemical plant complex

In 1915, Calco Chemical built a plant to manufacture coal-tar intermediates adjacent to the Raritan River in Bound Brook.[29] Calco was quite successful making synthetic dyes until the 1929 recession. It was acquired by American Cyanamid, a large growing chemical company. Cyanamid used the site to produce chemical intermediates and numerous other compounds, including pharmaceuticals, dye and textile chemicals, rubber compounds and various chemical intermediates. The facility was remarkably successful, operating 24–7 for quite a while. Cyanamid had ample land, access to clean water, rail and road access to the growing metropolitan areas to the north in New York and to the south in Pennsylvania. The surrounding area had a massive population, many looking for jobs in a time of economic distress. And Cyanamid's image was as an innovator. The fact that Rutgers and Princeton were nearby made the location even more desirable for a company that prided itself on scientific innovation.

American Cyanamid has been criticized for contaminating the area, and it certainly did that. Yet, in reality it was one of the most science-conscious companies, including about environmental science in the late twentieth century. Its location adjacent to a relatively small river in a state with at the time strong environmental protection focus led it to build a waste water treatment plant in

1940, later a large biological water treatment facility (largest in the New Jersey at that time), and in 1997 built a tertiary treatment facility with continuous activated carbon said to be the largest in the world at that time.[29,30]

Furthermore, Cyanamid invested heavily in process engineering bringing in engineering students to think about better ways of producing products, including fewer emissions.[30] Cyanamid figured out effective ways of managing aniline wastes. However, it could not manage all the wastes it produced from the massive operations, and this waste ended up in the river and/or was placed in large lagoons (see Text box 3.3). By the late 1970s, in addition to growing competition, Cyanamid could not escape the environmental legacy it had created, despite the reality that it was following accepted disposal practices before the Clean Water Act, the Safe Drinking Water Act, and the CERCLA legislation (Superfund).[31-36]

The following text box is from my notes of April 1977 (see Figure 3.5)

Text box 3.3 Run along the tow path, April 1977

What a crazy experience! I ran down the tow path separating the D&R (Delaware and Raritan) Canal from the Raritan River. After passing South Bound Brook, I got ready for the tree roots near the Cyanamid outfall. The discharge from AC looked pretty clean but the wind was blowing in my direction, and I got dosed with a really bad smell from the lagoons – like my body was being shoe polished. I picked up the pace to get through the smell as soon as possible.

Off in the distance, I saw two men riding my way along the tow path. As they got closer I could see they were wearing Revolutionary War uniforms. They stopped because the tow path was narrow as we intersected. They unfortunately were going into the plume and one said: "That ... Stink!!" I suggested that they needed to pick up their pace for the next 100 yards and they would be through it.

Later I learned that every year (continuing today) re-enactors celebrate the battle of Bound Brook, which was what I had witnessed. You didn't get shoe polished 200 years ago at this location, but you might get shot.

American Cyanamid was one of the 100 largest manufacturing businesses in the United States when I went on that run. It had over 100,000 employees, including 3000 at this site, and had developed tetracycline, one of the wonder drugs, among others. Over the years about 800 chemicals were produced on that Cyanamid site in about 100 buildings. One of my students learned as part of his thesis that the tertiary treatment sewage plant produced effluent that was cleaner than the water upstream and downstream of the site. But there was little forgiveness when it came to the costs of cleanup.

Meanwhile, the company began to sell off parts of its massive business to BASF, Pfizer, Proctor & Gamble and others. Those lagoons that I called the "shoe polish" lagoons, continued to be a problem for the company and the state environmental protection agency.

Cyanamid abandoned the site and it became a 575-acre Superfund site in 1983, one of the first national priority list sites in the United States.

Redevelopment of the site

Notably, 140 acres was not contaminated and beginning in 1996 that has been turned into The Bridgewater Promenade Shops (moderate sized retail mall), a baseball park for the Somerset Patriots (Independent baseball league, site was the parking lot for the Cyanamid site), and a New Jersey Transit parking lot. Meanwhile, most of the buildings were demolished by 2000. The remaining 435 acres were taken over by Wyeth (a subsidiary of Pfizer), which assumed the responsibility for remediating the site. By about 2000, the site was considered relatively "clean" according to various government reports, with one major exception.[31–33]

The major challenge was dealing with the lagoons/impoundments.[34,35] Reports showed 27 impoundments, which needed to be excavated. Some had corrosive and volatile waste for more than 50 years in lagoons used until 1981 and unlined for much of the time. Federal government[36] studies suggest that human exposures through water likely occurred since at least the 1930s. Pumping and treating water drawn from wells at the perimeter of the site restricted off-site migration, but not into the Raritan River. Waste accumulated over the synthetic liners and odors were common enough as my recollection of being shoe polished of 40 years ago suggest. The managers needed to pump out the material (estimated eight to 10 feet deep) and solidify it for treatment, thereby preventing it from ending up polluting the river. The lagoons are adjacent to Conrail freight line tracks (40–50 trains a day), which prevent continuous access to the lagoons. The estimated cleanup cost is about $200 million. Part of the risk management program has been to prevent trespasser access to the lagoons, which was reported in studies.[32–34]

In 2010, company representatives reported seepage along the banks of the Raritan River with measurable levels of benzene. They installed a groundwater removal system, a collection trench, a contaminant wall, and other engineered systems to halt the migration. This collection process will continue into the foreseeable future.

In March 2013, the U.S. EPA became the lead agency on the cleanup. In 2015, Wyeth agreed to spend $194 million as its part of the cleanup.[32] The fate of the 400+ acres is unclear. Various media and public documents talk about an industrial park, offices, others emphasize open space, and sports fields. Given the history of the site, including the potential for flooding and the legacy contamination issues, it would be hard to envision intensive uses. However, the large site might be able to support some business on the less contaminated areas of the site. The most likely result is open space that will allow walking, fishing, boating, and biking on the parts not impacted by those shoe-polish lagoons.

Overall, in essence, a major pharmaceutical industry selected this large flood-plain as a place to invent and produce its drugs. It was a major mistake, one guided by industrial location theory of the time. At least this company tried to install the best technology and even better technology than what was expected. Nevertheless, they could not figure out how to manage some of the wastes and suffered economically and politically as the impact of their site became known and better understood.

Somerset county cluster: South Bound Brook, asbestos roofing plant

The jurisdiction

South Bound Brook (SBB) is a 0.76 square mile (1.96 km²) borough with about 4500 residents in 2015 (see Figures 3.1 and 3.5). The case study is in this book because it represents a neighborhood-sized jurisdiction along a river with an adjacent canal that was heavily committed to manufacturing as we would have expected of a location with those physical attributes during the early twentieth century.

Under Bound Brook, I described meeting re-enacters of a Revolutionary War battle on the tow path across the river from the American Cyanamid site. That battle took place in what is now South Bound Brook, which makes the small river town a tourist destination. It was an outpost of 500–1000 men for the Continental Army under General George Washington (located in Morristown, NJ). A bridge across the canal and river first built in 1767 was the scene of major fighting. Plaques, structures, and various other artifacts are available to be visited, if you can get to them, and as noted above re-enactments occur every year.[37] The bridge is appropriately called the Queen's Bridge.

The area also has a direct link to the tow path that runs along the large Delaware and Raritan Canal (DRRC). The tow path is an excellent place for walking, biking, jogging, canoeing, and fishing. The author has often seen people fishing.

SBB is directly across the river from Bound Brook. Like Bound Brook and Finderne a large portion of the population is Latino, indeed most of the people I have seen fishing are Latino.

The GAF roofing plant

In 1895, a plant began to manufacture roofing products in South Bound Brook adjacent to the Queen's bridge along the canal. It was a logical decision. There was abundant fresh water, water and rail transportation, and an expanding market for roofing shingles. GAF could serve a large market between Philadelphia and its suburbs and New York City and its suburbs, as well New Jersey from this centrally located site. GAF built 15 structures on this site, which is literally a stone's throw away from Revolutionary war battles and the Continental army's centuries earlier outpost.

Some of the GAF buildings were made of brick, others of metal, and then there was a 200 foot high smokestack that towered over the landscape making sure that everyone knew what was there. In 1984, GAF closed the facility, along with 825 jobs. As much as the loss of jobs and tax revenues hurt, the legacy was, I believe, a bigger issue.

Here is an excerpt from my October 1974 notes about the site (Text box 3.4)

Text box 3.4 Observations about GAF plant, October 1974

Bruce (Newling) is running for mayor of New Brunswick claiming that asbestos from this site is polluting the water supply of New Brunswick. I think Bruce's claim is far-fetched, even though he is a friend and former colleague. After my run today, I am more convinced that this place has be knocked down and cleaned up. It is one of the ugliest factory complexes I have ever seen and this little town needs this land cleared to increase local tourism to the Revolutionary War sites a few feet away.

It really annoys me that the company seems to have been storing waste, which looks like deteriorated shingles on the tow path. There's a fence around it, but could easily be vandalized leading to asbestos exposure. This place is an embarrassment. It looks like it was hit by a bomb and no one came in to tear it down. It completely destroys the value of the adjacent residential area.

As it deteriorated, the site truly became an eyesore. In 1985, a voluntary effort to redevelop the site failed. In 1998, the Borough of South Bound Brook designated the site as part of a redevelopment area. In 2000, an environmental evaluation was conducted.[38–41]

Asbestos from shingle manufacturing was a critical issue at the site, a reality demonstrated by data reported by the company that identified and removed the asbestos.[38] They reported removing the following:

- 300,000 ft^2 of asbestos roofing and siding panels;
- 100,000 ft^2 of asbestos roofing;
- 11,000 ft of asbestos piping;
- 15,000 ft^2 of asbestos tiles;
- three five-story high industrial boilers with asbestos attached; and
- several hundred tons of asbestos contaminated soil.

In 2003, the 100-year-old 200 foot high smokestack was demolished as a visible sign that the industrial era in this small municipality was receding. As I noted in my text box notes from 1974, this industrial facility was the ugliest industrial facility I have ever seen. When that dilapidated mess came down, I cheered, and I can only imagine how happy all the people of the tiny town were. I volunteered to help the town review plans for redeveloping the site and was happy to do so.

Redevelopment of the site

The 15 structures were to be replaced by 152 townhouses. I can recall thinking that this might not be a good idea because the area was immediately adjacent to the tow path and the Raritan River was not too far away. However, the record of flooding here was not nearly as bad as across the river in Bound Brook and one of the reasons is that this area has a higher elevation. The GAF brownfield area was replaced by the Canal Crossing housing development. The units have two or three bedrooms and two or three bathrooms, between 1050 ft² and 2050 ft², and in 2016 when I consulted house for sale prices these were selling for between $215,000 and $350,000, depending upon the units' attributes.

To enhance this investment the area along Main Street that fronts Canal Crossing received a new streetscape including streetlights and new decorative pavers. In essence, realtors tell me that the units are selling to young married people who want access to the train station in Bound Brook, which is about a 10-minute walk from the house and development. In short, unless there is a very destructive flood, which has happened in this region in 1999, 2011, and 2012, South Bound Brook has found a comfortable transition from the industrial smokestack era to a transit oriented development.

Somerset county cluster: Manville borough asbestos and creosote plants

The jurisdiction

Manville is the furthest upstream among the six jurisdictions, about 17 miles as the crow flies from Perth Amboy and the New York Bight and 22 miles by auto (Figure 3.1). But Manville had the highest proportion of its labor force in manufacturing among the six: 69% in 1960. This city of 10,400 people and 2.4 square miles was the most impacted by manufacturing. Its name is inexorably tied to the Johns Manville Corporation for which the borough was named, and now is headquartered in Denver, Colorado. For many decades the company was the world's largest producer of asbestos products. The last time I visited Denver, my hotel was a five-minute walk from these headquarters. I remember wondering how many people working in that building had ever visited Manville, New Jersey, where the painful legacy has not been overcome.

Like Bound Brook and Finderne, Manville has had severe flooding problems. It was founded at the confluence where the smaller Millstone River empties into the larger Raritan River. Called the Lost Valley, the area at the confluence between the two rivers is problematic because of floods. The locational disadvantage is increased because the rail lines are situated higher than the homes creating a bowl-like configuration (see Figure 3.6). Figure 3.7 is a hand drawn picture of Manville at full production, drawn by the author's wife. It shows the massive size of the site and the utter domination of the landscape packed by asbestos manufacturing sheds over about 200 acres and connected

Figure 3.6 Manville sites.

Figure 3.7 Manville site drawing.

to the nearby river with a massive 48 inch pipe that symbolized how dominant the facility was.

Also confounding the difficult problems of redeveloping the area north of Main Street was a coal tar wood preservative plant that operated from 1911 to the mid-1950s. It severely contaminated the soil and local groundwater. Unfortunately, before cleanup, a developer purchased the site and built 137 houses in a small shopping area on that land. It lies immediately south of the Johns Manville site. The creosote seeped through the walls and foundations of the buildings. While the massive Johns Manville site had a much more severe impact due to direct worker exposure, the creosote site was a massive unsightly facility that needed to be properly remediated. EPA took over management of the site and spent $250 million of public funds to clean up a polluted area of up to 50 acres.[43] I was reviewer of the proposed remedy, and my concluding note to myself was "What an inexcusable mess!"

Over 400,000 tons of contaminated soil was removed, families were relocated, and businesses in the area were closed. It is no longer considered a Superfund site. Nor, however, is it likely to be redeveloped for any land use that can expose people through direct contact or groundwater. Currently, a pizza parlor, restaurant, a bank and a donut-coffee shop border the undeveloped site. The open space between these stores and the residential development is conspicuously unoccupied.

The Johns Manville asbestos complex

In 1912, Johns Manville picked a rural site near the confluence of the two rivers, three railroad lines, and a growing network of roads, plus a large local market for an asbestos manufacturing plant.[42] Asbestos is a hydrated silicate mineral that can be spun into soft silky fibers that have the property of withstanding heat and thereby function as a fire retardant.[44] The company began manufacturing cloth insulation, insulation board, pipes, roofing, rope, shingles, and various other asbestos-related products. Asbestos was brought from the company's mines, from Canada as well as from Russia and South Africa. The site covered about 200 acres and included 11 production facilities, an administrative building, warehouses, a power plant, and other smaller buildings. Large pipes delivered water to the site from the Raritan River located on the edge of the site. Waste was dumped at various locations in the borough (Figure 3.6). The historical pictures do not do justice to the site. Hence, my wife drew a picture that represents a composite of photos and my recollections, which a former colleague described as befitting an old-style prisoner of war camp.

Considerable debate has focused on what the company knew and did not know about the dangers associated with asbestos. Now it is a known carcinogen. Residents of the area recall white snowflakes wafting through the air and workers leaving the plant were labeled "snowmen."[42,45] For context, the author's father worked at the Brooklyn Navy Yard during World War II and was routinely exposed to asbestos as it was stuffed into the ships as a fire retardant.[46-47] My father's recollections supported by the literature are that men who smoked (he estimated half of the men smoked at work) were much more likely to die than those who did not. He was not a smoker and lived into his 89th year.

Many of those exposed were much less fortunate. By 1982, over 16,000 lawsuits had been filed against Johns Manville, which filed for bankruptcy. Friedman[42] reported that Manville residents had mixed feelings. On the one hand, the company donated land to the town, paid for various components of the infrastructure and paid relatively high wages and benefits. Furthermore, at its peak during World War II, 4500 workers were employed at the site.

However, as more investigations were done health impacts of asbestos became more obvious. Berry,[47] an employee of the New Jersey Department of Health and Human Services, led a team of researchers in an attempt to assess the risk of Manville residents who did not work at the company. The data show that asbestos-related diseases among non-working residents were ten times more likely than non-residents of the area.

Extraordinarily costly and painful lessons have been learned from the Johns-Manville case. Writing for the *Harvard Business Review*, Sells,[48] former manager for Johns Manville for more than 30 years, stated that managers did not believe the evidence of long-term consequences and characterized their behavior as a "colossal corporate blunder of the 20th century."

Redevelopment of the site

The former Johns Manville site has been redeveloped.[49-51] A major 12-theater complex was built, an adjacent retail mall, a Wal-Mart Supercenter, and small restaurants occupy the area along Main Street. The vast majority of the large site is occupied by ADESA, a large car auction company. I characterize the former Johns Manville area as a massive slab of concrete covered almost entirely by autos for auction that dwarf the space occupied by the movie theater complex, the Wal-Mart and a few other stores. Directly across the Raritan River from this several hundred acres of concrete is green open space.

Lessons learned

I cannot say with certainty that I know why sites in these six communities were picked, nor why they closed. I have no "smoking gun" evidence. With that caveat noted, if I had been in a management seat in the late nineteenth and early twentieth centuries, here are the traditional economic factors that would have attracted my interest to these sites for manufacturing and waste management:

- Minimize transportation costs: Access to water, rail and a road network so that raw materials and final products could be moved freely and inexpensively;
- Minimize transportation costs: Access to a massive local and regional markets in New York City, New Jersey, City of Philadelphia [10% of the national population in 1915];

- Minimize transportation costs: Access to nearby places, usually wetlands, where waste products could be buried, burned or emitted into the air with few restrictions;
- Labor and skills: Access to skilled and dedicated workforces;
- Labor and skills: A network of skilled inventors who found this region a near ideal place to develop and test new ideas;
- Infrastructure: Water, wastewater, electricity, other infrastructure were in place or nearby; and
- Local government support: Local governments welcomed these facilities as a source of jobs, taxes, and other community benefits.

By 1970, this traditional economic world was changing and the change was rapid after 1980, leading to a changing business conditions in these six and thousands of other places in the United States and Europe. Globalization of many resources, technology, communications, finance, and leaps in the ability to transport materials a long distance were major stimulants to change. New markets altered the center of gravity for many companies, as did a much larger and much less organized labor pool outside North America and Europe. The U.S. dollar strengthened relative to competitors in the 1980s and made U.S. products more expensive. Concerns about environmental and public health impacts of these industries, advances in environmental health sciences, and equity issues rose from issues that could not be ignored, especially by noxious industry and waste management facilities.

References

1 Hughes J, Seneca J. *New Jersey's Post Suburban Economy*. New Brunswick, NJ, Rutgers University Press, 2014.
2 Harris C. A Functional Classification of Cities in the United States. *The Geographical Review*. 33, 86–99, 1943.
3 Quitno Morgan. 13th Annual Safest and Most Dangerous Cities: Top and Bottom 25 Cities. https://web.archive.org/web/200801050957. Accessed August 26, 2016.
4 *Money Magazine*. 100 Best Places to Live 2006. #28 Edison. https://web.archive.org/web/20100802071714. Accessed August 26, 2016.
5 *Money Magazine*. Best Places to Live 2008: New Jersey. https://web.archive.org/web/20131203011336. Accessed August 26, 2016.
6 Mullins L. America's 10 Best Places to Grow Up. *U.S. News & World Report*. August 19, 2009. August 26, 2016.
7 Ford Motor Company. Ford Edgewater Assembly Plant. 2016. http://fordmotorhistory.com/factories/edgewater/index.php. Accessed March 27, 2018.
8 Stodhill R. Decades after a Plant Closes, Waste Remains. *New York Times*. July 29, 2007. www.nytimes.com/2007/07/29/business/yourmoney/29spill.html. Accessed March 27, 2018.
9 Isidore C. Ford Cutting 35,000 Jobs. January 11, 2002. http://money.cc.com/2002/01/11/companies.ford/. Accessed August 31, 2016
10 Fisher J, Hanley R. With Last 50 Pickups, Ford Ends 56 Years of Work in Edison, 2004. *New York Times*. www.nytimes.com/2004/02/27/nyregion/with-last-50-pickups-ford-ends-56-years-of-work-in-edison.html. Accessed March 27, 2018.

11 Rodriguez J. Engineering Company Can't Leave Suit Over Ford Plant Cleanup. www. law360.com/articles/375275/engineering-co-can't-leave-suit-over-ford-plant-cleanup. Accessed August 31, 2016

12 Hudgins M Turning Idleness into Opportunity. 2007. www.nreionline.com/develop ment/turning-idleness-opportunity. Accessed March 27, 2018.

13 Greenberg M, Anderson R. *Hazardous Waste Sites: the Credibility Gap*. New Brunswick, NJ, Center for Urban Policy Research, 1984.

14 U.S. Environmental Protection Agency. 1976. Kin-Buc Landfill Investigation. 2.27/76–3/23/76. nepis.epa.gov/Exe/ZyPURL.cgi?Dockey=9100GT0V.TXT. Accessed September 6, 2016.

15 U.S. Environmental Protection Agency. Superfund Record of Decision: Kin-Buc Landfill, NJ. EPA/ROD/R02–88/088/September 1988. https://nepis.epa.gov/Exe/ ZyPURL.cgi?Docket=9100SAUJ.TX. Accessed September 6, 2016.

16 Caroom E. Perth Amboy Refinery to Get New Life from $200 Million Overhaul. *The Star Ledger*. August 8, 2012. www.nj.com/business/index.ssf/2012/08/perth-amboy-refinery-to-get-ne.html. Accessed September 1, 2016.

17 McGurty J. Buckeye Says NJ Terminal Deal Gives Access to Water. February 10, 2012. www.reuters.com/article/buckeye-pipeline-perthamboy/update-1-buckeye-says-nj-terminal-deal-gives-access-to-water-idUSL2E8DA8AC20120210. Accessed March 27, 2018.

18 Gross J. Copper – The Evolution of an Industry. 2009. http://jegross.com/copper_ the_evolution_of_an_industry.htm. Accessed March 27, 2018.

19 Greenhouse S. Mini-Mills: Steel's Bright Star. *New York Times*. February 24, 1984. www.nytimes.com/1984/02/24/business/mini-mills-steel-s-bright-star.html. Accessed March 27, 2018.

20 Haydon T. Perth Amboy Steel Plant Shuts After 30 Years. 2009. www.nj.com/news/ index.ssf/2009/06/perth-amboy-steel-plant-shuts.html. Accessed September 1, 2016.

21 Rispoli M. Gerdau Announces Perth Amboy Steel Plant Closing, Sayreville Facility Suspending Operations. *The Star Ledger*. June 9, 2009. www.nj.com/news/local/index. ssf/2009/06/gerdau-announces-perth-ambly-s.html. Accessed September 1, 2016.

22 Loyer S. Gerdau's Sayreville Steel Mill Continues to Thrive. 2014. www.mycentral jersey.com/story/news/local/middlesex-county/2014/08/31/gerdau-sayreville-mill/14811925/. Accessed March 27, 2018.

23 Peschiera L, Freiherr F. Disposal of Titanium Pigment Process Wastes. *Journal of Water Pollution Control Federation*. 40, 1, 127–131, 1968.

24 Lynd L. Titanium. Econoatres.minyb1976v1.llynd.pdf. Accessed August 30, 2016.

25 Carney L. Environews. *New York Times*, August 15, 1982. www.nytimes.com/1982/ 05/23/nyregion/environews.html. Accessed August 31, 2016.

26 Carney L. Environews. *New York Times*, May 23, 1982. www.nytimes.com/ 1982/08/153/nyregion/environews.html. Accessed August 31, 2016.

27 United States District Court, D. New Jersey, Raritan Baykeeper, Inc. et al., Plaintiffs v. NL Industries, Inc. et al. Defendants. Civil Action No. 09–4117 (JAP), May 26, 2010.

28 Rothman C. Closings Begin on National Lead Redevelopment Site in Sayreville. *The Star-Ledger*. October 15, 2008. www.nj.com/news/index.ssf/2008/10/closings_begin_ on_national-lead.html. Accessed August 31, 2016.

29 Travis A. Dyes Made in America 1915–1980. *The Calco Chemical Company, American Cyanamid and the Raritan River*. Jeremy Mills Publishing, 2004.

30 American Cyanamid History. From International Directory of Company Histories.

Vol. 8, 1994. St. James Press. www.fundinguniverse.com/company-histories/American-cyanamid-history. Accessed March 27, 2018.

31 U.S. EPA. EPA Bound Brook Program: American Cyanamid Company, Bound Brook, NJ. https://cumulis.epa.gov/supercpad/cursites/csitinfo.cfm?id=0200144. Accessed September 7, 2016.

32 Palmer E. EPA Presents Pfizer with $194M Bill to Cleanup Former American Cyanamid Site. 2015. www.fiercepharma.com/manufacturing/epa-presents-pfizer-194m-bill-to-clean-up-former-American-cyanamid-site. Accessed March 27, 2018.

33 O'Brien W. Expect Heavy Machinery, Smoke from Contaminated Waste at Cyanamid Site in Bridgewater. N.J. 2013a. www.nj.com/somerset/index.ssf/2013/10/epa_pfizer_to_explain_detox_activity_at_American_cyanamid_site_in_bridgewater.html. Accessed March 27, 2018.

34 O'Brien W. Work Starting to Clean up Contaminated Cyanamid Site. 2013b. www.nj.com/somerset/index.ssf/2013/08/work_begins_on_American_cyanamid_superfund_site_epa_sets_fall_public_info_session.html. Accessed March 27, 2018.

35 NJ Department of Health. Revised Site Review and Update American Cyanamid Company Bound Brook, Somerset County, New Jersey, Cerclis No. NJD002173276. www.state.nj.us/health/ceohs/documents/eohap/haz_sites/somerset/bridgewater_twp/American_cyan/amcyan.pdf. Accessed March 28, 2018.

36 Winegar J. Revised Site Review and Update. American Cyanamid Company, Bound Brook, Somerset County, New Jersey, Cerclis No. NJD002173276. Prepared by NJ Department of Health and Agency for Toxic Substances and Disease Registry. 1994. www.state.nj.us/health/ceohs/documents/eohap/haz_sites/somerset/bridgewater_twp/American_cyan/amcyan.pdf. Accessed March 28, 2018.

37 South Bound Book in Somerset County Canal Crossing New Year's Sellout. November 13, 2006. http://njtod.org/396. Accessed September 7, 2016.

38 MACK Group. GAF-South bound Brook Asbestos Report Project. www.mackgrp.com. Accessed September 7, 2016

39 *The Chronicle.* GAF Site Will Come Crumbling Down. April 17, 2004. www.southplainfield.lib.nj.us/newspaper/chronicle/2004/2002–17.pdf. Accessed September 7, 2016.

40 *New York Times,* GAF Plans to Close 3 Roofing Facilities. January 5, 1984. www.nytimes.com/1984/01/05/business/gaf-plans-to-close-3-roofing-factories.html. Accessed March 27, 2018.

41 NJDEP. Former GAF Corporation Manufacturing Plant Site. Site Remediation Program. www.nj.gov/dep/srp/brownfields/success/gaf_sbb. Accessed September 7, 2016.

42 Friedman S. The Town Manville Built Has Mixed Feelings; the Talk of Manville. 1982. www.nytimes.com/1982/09/01/nyregion/the-town-manville-built-has-mixed-feelings-the-talk-of-manville.html. Accessed March 27, 2018.

43 EPA Superfund Program. Federal Creosote, Manville New Jersey. 2016. https://cumulis.epa.gov/supercpad/cursites/csitinfo.cfm?id=0204097. Accessed March 27, 2018.

44 Speil S, Leineweber J. Asbestos Minerals in Modern Technology. *Environmental Research.* 2(3), 166–208, 1969.

45 Hyman V. Roots Run Deep in Former Factory Town of Manville, NJ. 2011. www.nj.com/news/index.ssf/2011/03/roots_run_deep_in_former_facto.html. Accessed March 27, 2018.

46 Lilienfeld D. The Silence: the Asbestos Industry and Early Occupational Cancer Research – A Case Study. *American Journal of Public Health.* 81, 791–800, 1991.

47 Berry M. Mesothelioma Incidence and Community Asbestos Exposure. *Environmental Research*. 75, 34–40, 1997.
48 Sells B. What Asbestos Taught Me About Managing Risk. *Harvard Business Review*. March–April, 76–90, 1994.
49 Associated Press. Manville Closing Flagship Factory After 74 Years. 1986. www.nj.com/news/index.ssf/2011/03/roots_run_deep_in_former_facto.html. Accessed March 27, 2018.
50 Epstein S. $90.5 Million Awarded to 11 Manville Families Who Lost Family Members to Asbestos-Related Cancer. 2014. www.nj.com/middlesex/index.ssf/2014/07/905_million_awarded_to_11_manville_families_who_lost_members_to_asbestos-related_cancer.html. Accessed March 27, 2018.
51 Levy Konigsberg, Inc. Johns-Manville Asbestos Legacy in New Jersey. www.levylaw.com/johns-manville-asbestos-new-jersey. Accessed September 19, 2016.

Part II

Locating noxious facilities in the early twenty-first century

4 Redefining factors for locating noxious facilities

This chapter has three objectives:

- Consider the impact of human health and environmental sciences; laws and regulations; public perceptions, values, and environmental justice; and local government autonomy on siting noxious facilities;
- Note the emergence of and speculate about the implications of digital manufacturing and sustainability on siting noxious facilities; and
- Introduce a refined location process that is more inclusive of siting challenges in the early twenty-first century, and explain how it joins traditional economic location analysis and risk analysis to focus on specific sites and growing uncertainty.

Adjusting to new factors in siting noxious facilities

The importance of traditional economic location factors as described in Chapter 2 has not diminished. Scholars continue to revisit and expand the economics of location literature.[1-13] Making money and minimizing costs remain bottom line factors when choosing siting and non-siting options. Yet, powerful intersecting forces challenge the deterministic industrial location analysis paradigm that held sway for more than 50 years ago, especially for noxious industry and waste management facilities.

Environmental protection

By the early 1970s, it was clear to me that new laws and regulations in the United and soon thereafter in Europe were impacting the process of locating noxious industries. Anyone proposing a new nuclear power plant, landfill, large chemical complex, metal refinery, incinerator or other noxious facility was going to face the kind of government scrutiny, permit requirements, and other legal requirements that were much more intrusive than in the past. The question was how intrusive?

Part of the impetus was that the public expected their government representatives to protect them against what television told them was happening, what they read, and personally witnessed. The United States President Richard Nixon has been characterized as not particularly interested in environmental protection.[14] However, Perlstein also reports that President Nixon was concerned that he might lose the 1972 election to Senator Edmund Muskie of Maine or Henry Jackson of Washington if he did not support environmental protection. Hence, Richard Nixon signed the National Environmental Policy Act of 1970, many other environmental laws described below and appointed Russell Train as the administrator of the Council on Environmental Quality and later as second administrator of the EPA.[15] I was shocked when President Nixon took these actions, many of which impacted industrial and waste sites.

One of the first challenges the new federal agency faced was environmental cancer and its links to noxious industries. Pressed to develop a scientifically defensible way of classifying chemical agents, the EPA created risk assessment.[16] Using human epidemiological studies, animal tests, and chemical structure analyses, the EPA began to examine hundreds of chemicals in commerce. None of these studies were easy to do. There were huge data gaps in human databases, and animal tests proved to be less robust than had been hoped.[16,17] Some of these chemicals were found to cause or promote cancer, and others were found not to. For example, TaB, a soda with the artificial sweetener saccharin as an ingredient, was withdrawn when lab rats developed bladder cancer. The substance was later declared safe, and the drink is now back in circulation, albeit with a great deal of aspartame in it along with saccharin.[17] Risk assessment of hundreds of chemicals represented a level of scrutiny not previously practiced.

Gradually, searchable databases were built to make the research available. For example, the U.S. EPA created the Integrated Risk Information System (IRIS), which used risk assessment to classify chemical agents. The National Library of Medicine created the International Toxicity Estimates for Risk (ITER), which includes data from all over the world. These systems are easy to access on the web. They are never exactly up to date, and the information is often criticized. But like the TRI data (Chapter 1), these searchable databases represent a major improvement in transparency.[17]

The databases have been joined by amazing improvements in the ability to detect contaminants in the air, water and soil. Monitoring equipment became smaller, more potable and could detect concentrations of parts per billion and trillion, which are the equivalent of a penny in $10 million and $10 billion, respectively. The additional capability meant that it was possible to detect miniscule levels of toxins in many places. Jasanoff and Nelkin[18] point out that despite increasing precision we lack scientific knowledge about the hazardousness of many agents. For a large part of the public, however, finding any hazard meant they were going to contract cancer or some other dreaded illness. Chapter 1 summarized the creation of the Toxic Release Inventory (TRI), which collected data about many of these same chemicals, and then made it available to the public, which has made it more challenging for those that proposed sites.

Laws and regulations

In addition to the evolving science and databases, industry was squeezed by more than 50 federal laws and/or amendments passed between 1965 and 2016. Each of these directly or indirectly impacts siting of noxious industries in the United States. Table 4.1 lists 13 that I consider to be the most intrusive in regard to traditional industrial location theory practices (many more were passed – see the EPA website).

Each of these legal actions directly or indirectly impacted noxious industries. The bottom line is that after 1980 it was clear that industry, mostly noxious, could not emit waste in the cheapest way possible as had been assumed under Weber's and related industrial location theories (Chapter 2). Even with all the imperfections of the public databases, companies could not easily claim that their emissions had not been characterized and thereby posed no health and safety threats. Indeed, many companies used these data to try to find a substitute that was less hazardous (see Chapter 5). In regard to location, a company or federal agency would need to be foolish to ignore the reality that databases and mapping tools are now available to local governments and residents, as well as to special interest groups (see Chapter 9).

Some state governments followed the federal government's lead, indeed went further. New Jersey, as illustrated by Chapter 3, had some of the worst noxious industry problems. Governor James Florio of New Jersey said:

> Improper hazardous waste management is the most serious environmental problem facing our nation today.[19]

(p. 84)

In addition, fearing a high environmentally caused higher cancer rate, the State of New Jersey Department of Environmental Protection added science and legal staff, including a special group of environmental scientists who in short order began examining cancer patterns, the location of toxic emitters and the quality of ground and surface water drinking supplies. The state's population strongly supported a $100 million bond issue to provide resources to kick off these programs and it threatened to sue the federal government when the federal government delayed providing resources to manage chemical agents.

The bottom line is that the problems descried in Chapter 3 could have been diminished, perhaps entirely avoided or markedly reduced, with the kinds of policies and analytical tools now in play in North America, the European Union and increasingly elsewhere. Companies know that they may be prosecuted, sued by individual citizens, and their stock prices could be lowered. In regard to federal government agencies, illustrated by the Rocky Flats disposal case, when federal agents raided the U.S. DOE site and found illegal activities at a site about ten miles from downtown Denver, Colorado (Figure 4.1), public and private organizations need to avoid the risk of being on the wrong side of the law, as several of the cases in Chapter 3 demonstrated.

Table 4.1 Key legal actions that have impacted noxious industries in the United States

Law, Order or Regulation	Impact on noxious industries
Clean Air Act, 1963 and many amendments	Regulates air emissions; established air quality standards for areas that make location difficult in areas that do not meet standards; regulates hazardous air pollutants; requires many noxious industries to obtain permits to discharge
Clean Water Act, 1972 and amendments	Sets wastewater discharge standards; requires discharge permits from noxious facilities
Comprehensive Environmental Response, Compensation, and Liability Act (CERCLA), 1980	Established fund for cleanup of abandoned hazardous waste sites; placed tax on chemical industries (polluter pays principle); companies that purchased other companies were compelled to pay all or part of cleanup costs; updated to focus more on permanent remedies, encouraged community engagement in cleanup decisions
Emergency Planning and Community Right-to-Know Act of 1986	Requires industry to report on storage, use and discharge of hazardous substances to federal, state and local governments; required state and local government to prepare emergency response plans
Energy Policy Act of 2005	Focuses on reducing emissions of greenhouse gases, including industry and transportation; requires that biofuel be mixed with gasoline in the United States
Executive Order 12898 on Environmental Justice, 1994	Requires all U.S. government agencies to establish plans and processes that will reduce environmental exposures among selected minority populations and in poor communities
National Environmental Policy Act of 1969	First major law to establish the principle that environmental impact should be considered in federal government plans and actions; requires the preparation of environmental impact statements for federal projects or non-federal projects that require federal government permits and funding
North American Free Trade Agreement Implementation Act, 1993	Reduces restrictions on trade between Canada, Mexico, and the U.S.; impacted distribution and types of jobs, environment, and location of noxious industry in three countries
Nuclear Waste Policy Act of 1982 and amendments	Required a process to find and develop a single underground repository for defense and commercial nuclear waste; Yucca Mountain north of Las Vegas was selected, and is not open (Chapter 7).
Resource Conservation and Recovery Act, 1976 and amendments	Gives the U.S. EPA the authority to control hazardous wastes beginning with generation, including transportation, storage, treatment, and disposal; updated to include management of underground storage tanks
Safe Drinking Water Act, 1974 and amendments	Intended to protect drinking water supplies; amended to require assessment of cost and risk assessments when proposing changes
Toxic Substances Control Act, 1976 and amendments	As modified by Frank R. Lautenberg Chemical Safety for the twenty-first century Act of 2016, TSCA establishes data gathering, monitoring and risk assessment requirements for chemicals, including many already in commerce
Williams-Steiger Occupational Safety and Health Act, 1970	Created to protect worker health and safety, and established the Occupational Health and Safety Administration (OSHA)

* Source: U.S. EPA website is a good source for these. www.epa.gov/laws-regulations/summary-xxx.

Figure 4.1 Areas with noxious facility challenges.

Notice of what would follow from the federal laws came from government officials, business leaders and academics. John Quarles[20] was Deputy Administrator of the USEPA from 1972 to 1977 and EPA general counsel. In 1979, he wrote a 51-page essay in the *Environment Reporter* about federal regulations of new industrial plants targeted at chief executives, plant managers, corporate counsel, and other officers responsible for expanding industry. Quarles listed the Clean Air Act (various sections), the Clean Water Act, the Resource Conservation and Recovery Act, greater federal control of wetlands and coastal zones, new federal requirements for coal, the environmental impact statement process, among others (see Table 4.1). Quarles characterized federal legislation as representing an "awesome challenge to traditional practices of corporate decision-making, management and long-range planning…." He emphasized that the lead to time to plan and build a new plant would be "substantially increased," and any new plant with major emissions would face challenges he characterized as "profound"[20] (pp. 1–3).

A business perspective came from Richard Fortuna, then the Executive Director of the Hazardous Waste Treatment Council, an industry group addressing siting of a hazardous waste facilities in light of newly passed Amendments to the Resource Conservation and Recovery Act of 1984 (see Table 4.1). Fortuna[21] argued that the changes were the "most sweeping and ambitious that this nation, or any nation has ever attempted. It is the equivalent of our space program" (p. 1). He continued by asserting that society now realizes that the "market alone is an unreliable and indifferent broker when it comes to ensuring protective management of hazard wastes in a cost-competitive environment"[21]

(p. 4). Fortuna characterized this legislation as eliminating the option of care-less and deliberate pollution and the use of cost-based discretion by businesses. The focus of his remaining remarks was about the ending of the solution that would allow inexpensive and temporary land disposal (see the Kin-Buc case in Chapter 3). He added that the good side of this for business is that it provides a "sense of certainty" about what was required. Fortuna notes that hundreds of unlined and leaking landfills already had closed and others were being taken out of service because they could not certify that they could protect groundwater.

Howard Stafford,[22] a major figure in industrial geography, used his excellent access to business leaders to interview more than 50 business executives who located branch plants. Stafford, one of the developers of traditional industrial location theory, designed interviews so that respondents would comment on what he called "clean plants" and "less clean" ones. Less clean plants were chemicals, steel, engine manufacturing, and smelting operations. Across the full set of interviews, environmental regulations ranked 8th out of 10, about as important as taxes. However, I divided the interviews into two sets and was able to see that environmental regulations ranked 3rd in the less clean group – behind materials and markets, but ahead of transportation costs, site character-istics, utilities, investment costs, labor and business climate, for example. Stafford noted that a decade earlier, environmental regulations would not even have been on this list. Further insights from his study were that non-attainment areas for air pollution, older industrial places, and newer industrial areas in states California and Oregon were more difficult sites for less clean industry, a point also made by Quarles.[20]

In short, these three papers illustrate the increased complexity faced by noxious industry managers interested in siting a facility beginning in the 1970s.

Public perception, environmental justice, values and risk communications

Traditional location theory did not acknowledge public reactions. There were good reasons for this practice. First, industry brought jobs, income, and wealth to communities. Developers expected that local people would speak out in favor of new sites. Many local people supported the noxious industries described in Chapter 3, and when I was studying location theory I was told that people living in industrial neighborhoods would accept higher decibel levels because their jobs depended on accepting more decibels. The second reason to expect support or no negative public reaction is that the public has a lot more to worry about. Under normal circumstances presumably they would not focus on a nearby factory, unless there was a truly annoying smell, loud sounds, or notably distress-ing visual images.

As part of the post-1970 consciousness raising about the environment, researches began to accumulate evidence that the public was worried about exposure to carcinogens, toxins, and that even the economic benefits were less than had been expected and in some cases were negative, for example reducing property values.[23-32] Even though the results were mixed, that is, not every noxious facility was perceived or measured to be harmful, proponents of existing sites and new ones faced opposition that was new to them. Over the years, the siting of the nuclear waste site at Yucca Mountain in Nevada has become the poster boy for public challenge to siting. To some extent this failure has been offset by success in siting the Waste Isolation Pilot Plant near Carlsbad New Mexico (Figure 4.1) with public support[29] and progress toward sites in other countries (see Chapter 7).

Psychologists Paul Slovic, Dan Kahneman, Baruch Fischhoff, and others found that the public's collective reaction to possible risks is based on emotions, feelings, and images.[23-25,30,32] Rather than use so-called "rational" thinking processes, the public, they found, builds mental shortcuts, called heuristics, based on images, feelings, key words, and they rely on their heuristics and those that they know and trust, rather than on scientific data (see Chapter 1 for discussion of images and colors in classifying hazards). This literature distinguished more than two dozen factors that tend to cause greater public concern and those that reduce public concern.[32] I have listed nine that I have repeatedly seen distress people who speak out at public meetings:

- Involuntarily siting in a community;
- No local control of the facility;
- Unfamiliar chemical, physical or biological agents and technologies proposed;
- Proposed siting perceived to be unfair, for example, clustered in disadvantaged communities;
- Bad memories of the type of facility;
- Emissions from proposed site can lead to acute and fatal outcomes;
- Facility uses "artificial" substances and/or manufactures them;
- The proposal is "immoral" (e.g., emits to water sources);
- Developers provide false or misleading information.

My former colleague Peter Sandman[32] characterized these attributes as producing public "outrage." Large noxious facilities are easily branded with many of these negative labels, which the public stores and recovers as powerful images, colors and other labels that reduce the credibility of siting proposals and scare others into strong opposition. I have witnessed this happen dozens of times. Without doubt the toughest images in my experience are attached to nuclear and chemical waste management facilities (see Chapter 1 for some examples from one study).

Economic impact of noxious facilities on different populations is a prominent issue. Clark et al.[31] examined wage and land rent differences that result from environmental amenities and disamenities. Using 1980 census data, they examined economic impacts of 262 plants that generated electricity with nuclear, coal, gas, and oil as their fuels; military facilities; hazardous waste sites; petrochemical refineries; liquefied natural gas storage sites; and radioactive waste sites. Further subdividing the results by home owners and renters, they found stronger negative values for home owners than for renters, which should be expected given the economic investment of the owners, which is consistent with the literature (see also endnotes 27 and 28).

Whether the issue is lower property values, exposures from emissions, or another issue of public concern such as truck traffic, the underlying issue is trust. Rather than summarizing a long set of papers, I will illustrate with a case. In Anniston, Alabama (Figure 4.1), the U.S. Department of Defense was engaged in the process of advocating for an incinerator to destroy tons of nerve and mustard gas.[17] The author and colleagues were asked by the U.S. National Academy of Sciences to work on a risk analysis and meet with local community members. The community members were respectful, at first, but increasingly hostile, partly because none of us were from the local area, all but one was an older white male, all had PhDs and in other ways were not representative of the local population. The community brought distrust to the table. For example, they argued that the Army would bring in other industrial "trash" and burn it in this incinerator. When one of the most expert mechanical engineers in the world explained that the $800 million incinerator was designed specifically for liquid waste not solids, the residents were reluctant to accept his statements.

As the debate evolved, trust became the transparent issue. For example, I was challenged by a comment from members of the audience that the incinerator would lower their property values. They had found several papers that showed that hazardous waste sites and incinerators lowered property values. We also faced verbal attacks that we were paid consultants, otherwise known as "hired guns." I tried to explain who we were, where we were from, and that we were not paid a fee. I was initially met with hostility, which slightly decreased when other committee members described other public projects they had worked on. A large segment of audience was unwilling to trust outsiders to advise them about property values or other issues. The key issues were trust and credibility not science.[33–35] We were seen as not sharing their values, some of us communicated in jargon that they did not understand, but more important they charged that we would leave town and not bear the economic, health, and social consequences of the incinerator.

In the end, after much debate and law suits charging environmental injustice and a great deal of distrust, the plant was built. All of the weapons were destroyed and to the best of my knowledge no one was exposed to nerve or mustard gas.[17] Years later, I am told that some residents remain angry and distrustful. Unfortunately, it is a lot easier to understand trust than it is to build and especially to retain it.

Environmental injustice has fired a piercing arrow into the heart of some noxious facility siting proposals. The United Church of Christ (UCC) published a study in 1987 focusing on the location of 415 commercial hazardous waste facilities in the United States. With an acknowledgement that I was the technical consultant to the study, the major finding was that zip codes that had two or more hazardous waste facilities or one of the five worst hazardous waste facilities in the United States had an average of 36% minority population. Those that had one of the 415 hazardous waste facilities had 24% minorities. The last set of zip codes had no facilities and had an average of 12%.[35] Not surprisingly, poor residents disproportionately lived in these zip code areas with waste sites. This finding was for 1980 data, and then in 2007 the UCC published a new study with similar results as the earlier one.[36] Others have made findings that contradict these and some have confirmed these findings,[37-42] but the charge of environmental racism was heard by government officials from the President's office to local community groups and has made a big difference in regard to siting noxious facilities.

Also heard, perhaps even more so than the quantitative studies, were a series of book-length studies underscoring environmental justice across the United States. Authors examined segregation, the location of industry in areas zoned for Blacks and Latinos until this practice was declared unconstitutional. There was considerable debate about whether developers were deliberately proposing a noxious facility for a location because the population did not have the power to resist or whether the area was already industrialized and the proposed noxious facility was a continuation of current practice in the area.[43-48]

The most powerful of these in my experience was case studies in which the authors took great care in interviewing local residents who were passionate about their personal experience and their neighborhood. Steve Lerner's award-winning book is the latest of these case study books.[48] Lerner visited 12 places across the United States. Six sites had industries and the other six were waste management facilities, primarily U.S. government ones. The study areas were places with about 300 people living on the other side of a fence across from a major noxious facility or on top of or adjacent to a closed waste facility. Almost every one of these cases had the following attributes:

- A government or business charged with chronic emissions and periodic massive releases across the fence out to five miles from the site, creating noxious odors and fumes, lights and flares, and distressing sounds;
- Non-responsive local governments, at least at first;
- State governments that were slow to act or would not act;
- A few people reached out to the state and/or federal agencies for assistance;
- Local residents organized and gathered data from the Toxic Release Inventory (Chapter 1), and took their own environmental samples, as well as launched mass protests that attracted the attention of the responsible parties; and
- Court challenges were directed at reducing the exposure and/or relocating some or all of the exposed population.

Reading Lerner's book, it is easy to identify with the people that he interviewed. They seemed sincere, hard-working people who had been ignored by business and government. They stated the NAIMBY (not always in my backyard) acronym. I picked one quote from each of the first three chapters (see Figure 4.1 for locations). The very first sentence of the first chapter starts as follows:

Ruth Reed, resident of Ocala, Florida who lived next to a charcoal factory:

"I just got mad. I couldn't breathe in my own house. It smelled like lighter fluid."[48]

(p. 1)

Margaret Williams, resident of Pensacola, Florida who lived in an area between two factories (one treated wood with creosote (see Chapter 3 case in Manville); the second was a chemical fertilizer plant), said:

We thought they [EPA officials] were coming in to help us. This reckless-ness would not have occurred in nonminority or wealthy neighborhoods.[48]

(p. 49)

Hilton Kelly, resident of Port Arthur, Texas who lived 300–400 yards from a petrochemical complex on the west side of Port Arthur, Texas:

This area is not safe. We want to move so that we can get a chance to live.[48]

(p. 75)

Every chapter in this book has powerful personal stories to tell and the cumula-tive impact of the personal stories leaves a strong emotional imprint, even if you disagree with some of the facts and cause-and-effect assertions made in the book. This is perhaps the most effective of these books, but the others I listed leave strong emotional messages, irrespective of any science that is part of the story.[43-48]

Several environmental justice cases reached the U.S. federal courts. Plain-tiffs alleged that the Civil Rights Act of 1964 was violated by ignoring cumu-lative risk implications in Chester, Pennsylvania and Camden, New Jersey.[17] The author is familiar with both of these cases. A clear takeaway was that the businesses that proposed the siting believed that they were following tradi-tional industrial location practices. Smary,[49] an attorney with a business per-spective, characterized the environmental justice issue of the middle 1990s as a very emotional issue for all sides and one based not only on equity but on cumulative risk.

Over three decades after these environmental justice debates began, as a person involved in the emergence of this issue, I no longer raise the distinction between deliberate and historical siting as important. Any government or private manager that proposes a noxious facility would be out of touch with reality and threatening their reputation by not checking to determine the racial/ ethnic and income distribution of nearby populations. In fact, if I was heading a group proposing a noxious facility, I would examine the demographic attributes of population out to five miles as a high priority fact-finding item for any region that survived other screening tests. It is predictable that this issue will be raised, and Chapter 9 shows that there are readily available tools in the United States to make a finding about social and environmental justice. Not to avail oneself of publicly available data is not acceptable, even if you disagree with the data and/or the results it produces.

The field of risk communications has grown in response to public concerns about social and environmental justice. The initial idea was to craft convincing messages that would turn opposition into cooperation at public meetings and the media. The mechanism has been markedly less effective than had been hoped for when ideas was first touted.[50–52] For example, a brilliant speaker may have little credibility because s/he is seen as not representative of the community. Second, public meetings too often become a public free-for-all, with groups and individuals using the opportunity to state their views, cut off other speakers from stating their views, and in the process gaining publicity for their cause, whatever that might be. Interest groups on all sides of the issue have learned how to use public meetings to their advantage. Furthermore, multiple public meetings about the same topic may lead to different and contradictory outcomes. Different people attend different meetings, new information is introduced, hence it is not surprising that sometimes more opposition emerges, other times opposition softens.[53] If a large private or public organization is going to communicate, it needs to spend time developing objectives, a plan, and then be sure that the speakers are ready to talk and listen.[51]

Risk communication is a good step, but decision-makers must realize that when a debate is about values, there may be no real listening and changing of public opinion. Public meetings may be a good exercise in venting of public opinion, not changing it. If the goal is to modify public perceptions and preferences, decision-makers need to listen, which may begin at a public meeting, but not end there. A good follow-up to an initial meeting is breaking into smaller groups based on attendee interest to focus on specific issues with reporting back to a larger group. Even in the worst case situation where a plurality appears to be opposed to the proposed site, there are some people who will be in favor of it and many who are open-minded and be willing to listen (see Chapter 7).

Government

As a graduate student, I learned that most local governments wanted industry, even large plants because they delivered jobs, taxes, and sometimes indirect benefits to communities. The basic–nonbasic distinction (Chapter 2) underscored the importance attached to manufacturing. I believed, as did many of my professors and colleagues, that local government was the easiest most predictable supporter. Given that Harris (see Chapter 2) classified over 40% of U.S. cities as manufacturing, this means a great deal of public support, or did at that time.

Over a half century later, I would characterize local government officials as the most unpredictable part of the augmented set of location factors because of local autonomy issues and political processes that trigger their reactions. I have been involved in cases where local officials started and ended with the same position about a site. I have been involved in others when they changed their minds, without as best I could determine, any new scientific information. Then, there are cases where an election brought in a new mayor or the equivalent that opposed a proposed facility because that elected official had different knowledge and values than the prior one. The new mayor may have no investment in implementing a policy that will be credited to the prior mayor, especially if they ran against each other. A change can leave a private or government developer with a serious dilemma. Should they negotiate? Threaten a law suit? Cancel the project?

The Not in My Term of Office (NIMTOO) acronym brands this problem in the hide of local government. Lake[54] argued that local government needs to represent local interests against business and federal and state governments that all too often are protecting the power of capital and favoring mechanical-like standardized process over substance.

As Chapter 8 discusses, the same dilemma faces facility developers who want to locate a plant outside their national boundary. They may be successful, and they may also find themselves opposed by a new government that confiscates their investments, insists on renegotiations that makes the site infeasible, does not guarantee security, and in other ways is hostile.

Overall, the net effect of laws, regulations, risk perception, environmental justice, values, and the assertion of local autonomy has made proposed new noxious facilities a face-off of values in the minds of many people. Developers will need to be prepared to cope with the following simple juxtapositions about siting that will be asserted by the parties often in rude language accompanied by unflattering gestures, such as, but by no means limited to the choice between:

- Human health and the environment versus economic growth;
- Community autonomy versus the needs of capital;
- Local control versus federal and state pre-emption;
- Decisions based on values versus science;
- Soft and dispersed solutions versus hard and centralized technological solutions; and
- Fair share of the burden versus inequitable burden.

New considerations with uncertain impacts on siting: digital manufacturing and sustainability

With perfect 20–20 hindsight the consequences illustrated in Chapter 3 could have been entirely avoided in some cases or at least reduced. I like to believe that proactive planning, scanning and implementation is far more desirable than waiting and hoping to avoid hostile and costly reactions that resulted from the traditional economic location approach for siting. If I am right, we can accommodate the following powerful emerging trends during siting and non-siting stages: digital manufacturing and sustainability.

Digital manufacturing and sustainability appear to me to be gaining momentum toward reshaping manufacturing and waste management. While siting has not been their primary outcome, I briefly comment on them here, and note that they will reappear in the case study chapters, especially Chapter 5.

Digital manufacturing

Globalization and environmental concerns described above pose two major challenges for industry. The turn to intelligent machine manufacturing may be a near term game changer for siting.

The design of factories and some waste management facilities described in Chapter 3 had several distinct steps:

- An idea for a production process is prepared with hand drawings;
- Some sculpturing of the machines and tools is completed;
- A prototype is built;
- Architects and engineers use their experience and handbooks to draw blueprints;
- Machines and tools are manufactured by expert craftsmen;
- Raw materials are assembled;
- Machines and process are laid out on a floor(s) in an assembly line;
- Staff secure water, energy, telephone; transportation; other infrastructure and legally permissible way of managing wastes;
- Products are manufactured and stored in preparation for shipping;
- Some waste is recycled; and
- Some records are maintained, including by technicians who periodically check the machines and tools.

By the late twentieth century, computers, electronics, telephones, and spreadsheets were introduced and these allowed some creative alterations to manufacturing to be planned and implemented. Readers who lived through this period will recall how these changes impacted their work. When I began my research in the 1960s, I used a slide rule (my birthday present for going to college) and then a mechanical calculator. We began to use electronic calculators during the early 1970s. The electronic calculator was amazing. Those devices, along with my slide rule and mechanical calculator are now antiques that amuse my students. Today, I can do calculations on my tiny personal computer in a second or two that took two to three hours on the mainframe computer in the basement of the Columbia University engineering building, but only after I had punched all the data on to IBM cards and inserted them into the card reader.

Going from mechanical to electronic and now to digital manufacturing is going to transform manufacturing, even for noxious facilities. Today's so-called "smart" factory consists of connected machines and tools that continuously record and analyze data, and make changes. They can improve performance, reduce resource use and notably can spot flaws that if not dealt with can cause a serious problem that impacts human health, the environment, property values and community harmony.[55-60]

At the concept phase, hand sketches of a half century ago have been replaced by 3D sketches and models. Technicians use continuous data feeds to make changes. Part of the benefit is that less is wasted, including resources and part of the objective is to adjust the process so that wastes become raw materials for other industries. The case study literature reports the following benefits from digital manufacturing:

- More rapid adjustment to customer needs;
- More efficient and flexible processing;
- More reliable final products;
- Less resources needed;
- Less need to store material and final products; and
- Less space needed.

Reports tout digital manufacturing for the reasons listed above.[55-57] The automobile and aircraft industries have been major innovators, and the U.S. Department of Defense has been a major supporter of this science. I find less from the chemical, petrochemical, metal, paper, and nuclear industries but they are engaged.[58-60] For example, an executive in the paper industry told me that the paper industry has continuously adapted to changes, and that he believed that digital manufacturing would ultimately allow the industry to markedly cut down on its resource use and waste production, as well as manufacture precisely what customers wanted. A chemical plant manager told me that the sensors imbedded in the manufacturing process allowed the staff to fix problems before they could cause a serious exposure.

I have lived long enough to know that promises based on technology do not always come true. I list five challenges that must be addressed:

- Intellectual property issues are a major economic issue;
- Cyber security and vulnerability are daunting problems;
- International agreements are essential to make sure that users are measuring elements such as length, volume, voltage and other parameters in the same way;
- The technology sounds simple, but will be costly to develop and there will be setbacks. In this era some businesses and countries are notably reluctant to invest in the future; and
- Organizations need to have leaders that are willing to create an atmosphere that rewards creative scientists and engineers who are charged with building and enhancing the factory of the future.

Assuming digital manufacturing replaces twentieth century processes, what are the implications for noxious industry siting and waste management? While I do not have a crystal ball, I suggest that three different scenarios are likely. In the case where the plant is producing a final product that has the potential for virtually eliminating noxious materials and emissions, digital manufacturing because it requires less space and allows real-time adjustments negotiated between customers and producers will enjoy greater flexibility in siting. This includes portability that will allow production to be fit into smaller production facilities with less opposition and more local support.

Some facilities will use resources more efficiently than they have, but noxious substances will be apparent to those that search databases. For these, I expect more geographical concentration of noxious industry that can use digital manufacturing to gain greater economic and resource efficiencies, and be located in places that will allow waste to be turned into resources. Given globalization and the increasing size of water vessels and rail shipments, it follows that ocean ports and selected rail-oriented internal locations will be the most flexible and practical places for location, assuming that the facilities noxiousness can be reduced. Many potential sites already exist in the form of former smelter, refinery, gasification and other brownfield sites. Many of these remain and are economic burdens on companies and government because they will need remediation to a residential standard unless reused for manufacturing, storage, or transportation. The logical option is to use these sites, so-called "mothballed" sites for these purposes (see Chapter 10).

Third, even with digital manufacturing process, some facilities will continue to be obviously noxious, primarily because they use and/or create noxious materials. For these I see more remote locations and clusters in places with relatively few people and compatible land uses (see Chapters 6 and 7).

Sustainability

Quoting the U.S. Environmental Protection Agency, sustainability is

> to create and maintain conditions under which humans and nature can exist in productive harmony and that permit fulfilling social, economic, and other requirements of present and future generations.
>
> (NEPA 1969, Executive Order 13514, 2009)

There is a massive sustainability literature, much of takes a global perspective and is about urban areas. However, the idea of sustainable industrial development and waste management has certainly not escaped industry and government. Eric Johnson's book traces sustainability in 29 large chemical companies, 23 of which have a program. He describes sustainability as trendy, even among the "uncool" chemical industry.[61]

The paper industry is arguably even less cool than the chemical industry. Paper manufacturing was once fixed in forested areas, and later branched out in cities where rags and scrap paper became the source (Chapter 2). The industry has always faced environmental challenges from the days of the hydrogen sulfide air emissions to present urban locations where de-inking processes and fibers generate sludge that needs to be managed. Hence, the paper industry was one of the first to adapt pollution control and eco-efficiency measures to reduce costs and noxiousness. Sharma and Henriques[62] report that the industry has begun to cluster facilities in order to reduce waste and turn it into raw products for aligned industries. They add that companies are quickly moving toward greater sustainability by adopting procedures and technologies that substantially cut down on water and energy use, and replace chlorine in the manufacturing process.

A few years ago, I was approached by a company with suggestions for changing the paper used to print a journal. After a few phone calls, I was told by members of the industry association that that the industry has to be at or near the top of the list in adopting sustainability because its products are so easily traceable. It is more than just reducing emissions from factories and changing forest management practices, it is about investing in digital manufacturing and taking the sustainability pulse of customers.

In this regard, the Environmental Paper Assessment Tool (EPAT), a product of the industry's biggest manufacturers, illustrates the need for this industry to be cleaner and more sustainable. EPAT allows customers to assess different kinds of products on the basis of efficient use and conservation of raw materials, cleaner emissions, and impacts on surrounding communities. Suppliers and buyers subscribe to EPAT, and their performance is rated on 20 indicators normalized to industry averages and on a 0 to 10 scale. (There is a charge for suppliers and buyers to use the product.[63])

The federal government has required itself to invest in sustainability and to lead by example. Executive Order 13693 (Planning for Federal Sustainability – March 19, 2015) requires that federal agencies develop and implement a sustainability plan. The agency plans receive an annual rating from "green," which really essentially means good performance to "red," which means poor performance. The agencies publish these ratings. For example, in 2016 the U.S. DOE sustainability report card had green signs for emission reduction, reduction in energy intensity, use of renewable energy, reduction in potable water intensity, and reduction in fleet petroleum use. It received a "red" flag for green buildings, which means only 7.6% of buildings were sustainable. A review of the annual plans that begin in 2010 shows clear improvement in all of the categories, except the last one.[64,65]

Digging into the details behind the scores was useful. For example, the DOE environmental management program is in charge of numerous buildings that are scheduled to be demolished. Protecting them against collapse, fires, and other events makes sense, but spending taxpayer money to make them sustainable is illogical.

The DOE has, in fact, sited some new facilities with sustainability in mind. For example, at the Savannah River in South Carolina they built an energy plant that is fueled by wood chips from surrounding forest and tires located on the site. This plant replaced on old coal plant dating from the 1950s that was inefficient and created emissions that were considered unsatisfactory. The new plant was done to meet the intent of Executive Order 13423. Good plans, however, have some drawbacks. In this case, the wood plant could not cope with a polar vortex that hit the South Carolina area in 2014, and parts of the system froze, causing a power shortage for a short time. A new plant has been added to solve that problem. Second, I was told that the plant managers were criticized by those who argued that that building a locally sustainable wood burning plant was not part of the site's nuclear mission.

Given the fact that the public does not live close to any of the DOE sites, one might think that sustainability does not matter. However, that assumption would not be a correct one. In 2013, we surveyed 922 people living within 50 miles of one of the following U.S. DOE nuclear sites: Hanford, Idaho National Laboratory; Savannah River and the Waste Isolation Plant (see Figure 4.1).[64] Respondents were asked to rate the importance of 13 different DOE on-site activities. The highest possible score was 10 (most important) and the lowest was 1 (not at all important). The most important action was to regularly monitor the health of site workers (avg. 8.9), followed by scanning air and water quality at the site (avg. 8.7), and then providing training and equipment to emergency responders from surrounding regions. Rated 6th out of 13 was to protect on-site forest, soils and animals (avg. 8.3). Rated 10th was advancing the regional economy, a measure of economic sustainability (avg. 8), and the lowest ranking desired public action was that the DOE should reduce on-site energy, water and fuel use (avg. 6).

While the three sustainability measures were middle to low in attracting public interest, only 14% expressed no interest, and 5% said they were interested enough to serve on a community committee that interacts with DOE about on-site sustainability.

Will sustainability change siting practices? It is not clear that the abstract idea of a more sustainable noxious neighbor will make an important difference in of and by itself. What is more salient is that organizations are thinking about sustainability as they consider their ideas about siting and non-siting options (see Chapters 5 and 7).

A refined siting process for the twenty-first century

The post-1970 effort to improve air, water and land quality marked the third twentieth-century major environmental effort in the United States.[66] The first was the creation of the U.S. National Park system by President Theodore Roosevelt in the early twentieth century. During the 1930s, after major floods killed thousands of people, President Franklin Roosevelt took actions to manage floods, which was the second environmental effort. The third effort has focused on protecting human health and the environment beginning in 1970.

As part of that effort, laws and regulations, improvements in environmental science, and the introduction of perception and distributive justice has forced manufacturers, especially those that deal with noxious agents to adjust their practices. All of this has occurred while the world experienced unprecedented economic growth. From the perspective of the developer, Figure 4.2 summarizes where I think we are in regard to siting process for noxious facilities from the developer's perspective.

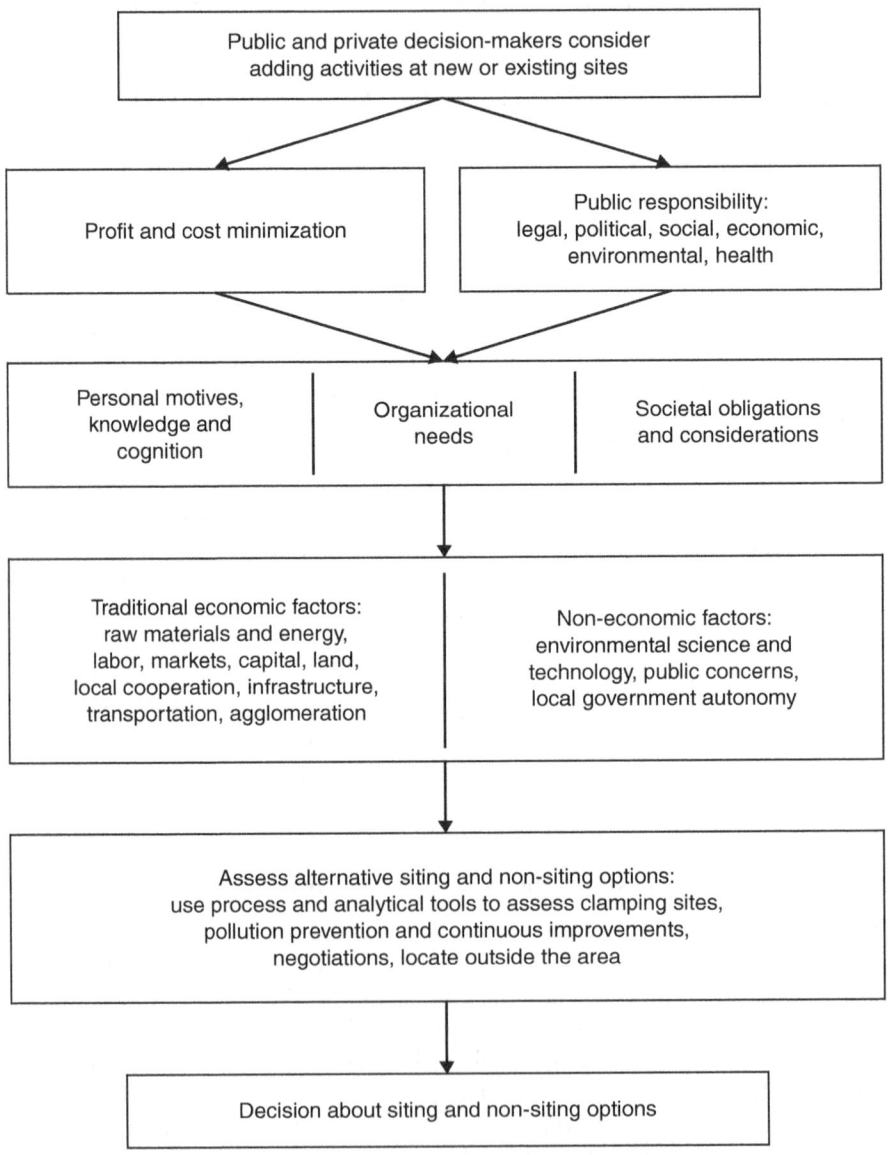

Figure 4.2 Summary of twenty-first century process model for noxious facility siting.

Before getting to the point of consulting traditional and newer siting factors, decision-makers should have examined their organizational needs, principles, and personal motivations. It may be that these principles and motivations have changed due to corporate mergers, personnel changes and so on (see Chapters 5 to 8).

Assuming that there is a preference to look for a site(s), then the challenge is to blend the traditional and the newer location factors described in Chapter 2 and earlier in this chapter. Chapters 5 to 8 describe four emerging trends that should be central to their considerations:

1   Becoming less noxious – avoiding actions that cause objectionable odors, sounds and distressing visual images and emit hazardous materials through pollution prevention and inherently safer designs – Chapter 5;
2   Choosing Locations at Major Plants (CLAMP) – rebuilding or expanding on sites that the organization controls – Chapter 6;
3   Negotiating – Engaging in discussions and bargaining with interest groups regarding the conditions for siting or non-siting solutions – Chapter 7;
4   Letting it Go – Deciding to move all or part of the manufacturing and waste management to another location or give up the production – Chapter 8;

To use a common expression, there are no free lunches in any of these options. Some could violate the organization's principles and established culture, cause delays, fail, and in the long run may be counterproductive in regard to making business, social, environmental or political sense. Each of the four case study chapters describes one of the above four strategies, its advantages and disadvantages, cites illustrations from the literature, and explores one or two illustrations in detail. Organizations in many cases will need to avail themselves of all four in order to cope with emerging realities, or at least muddle through (see Chapter 10).

A holistic framework that accommodates the breadth of information and viewpoints that need to be accommodated in the early twenty-first-century siting process would be a valuable asset for those who need to make sure that the myriad of layers involved in siting are captured and balanced. Three obvious options are as follows:

•   Environmental impact statement process (EIS);
•   Benefit and cost analysis (BCA); and
•   Risk analysis.

The first two provide vital material but are widely perceived as unbalanced. The EIS process produces masses of data about programs and specific site-related options. However, the information as the name suggests is heavily weighted toward environmental concerns, which it was designed for. Later revisions have brought some but limited information about economic issues and environmental justice. Elsewhere, I have written that the economic impact and environmental justice elements are almost always the weakest parts of EISs.[67] BCA has the same limitation in reverse, that is, human health and safety, and environmental data are present, but often lacking what is needed for a non-monetary assessment of the options. In reality, BCA or something like it is institutionalized and will remain a key framework for judging results, and the EIS process will remain central to large noxious facility projects because it is likely to be required.

Risk analysis as a framework also brings some baggage to the table. Some confuse it with risk assessment, which it incorporates, and many consider it to be a narrowly defined process for studying chemical as carcinogens, nuclear power plant safety and hazardous waste facilities. Those were the three first major uses of it, but it has expanded and has always has had the flexibility to accommodate siting issues with an emphasis on uncertainty.[17] I believe that a risk analysis framework can examine the following key questions that are part of the siting of noxious facilities:

1   What events can happen at a location or to a non-siting option that can substantially increase the benefits and decrease negative consequences for the organization and the area?

2   What are the chances that something with serious positive or negative consequences will occur with a siting or non-siting option? And when and under what conditions might these occur?

3   What are the consequences if something of major benefit or serious negative consequence does occur?

4   How can consequences be prevented or reduced in the case of a negative event, or enhanced in the case of a positive one?

5   How can recovery be enhanced, if the negative event occurs? And how can benefits be sustained in the event of a positive event?

6   How can the decision-makers be organized to be as responsive as soon as possible and responsible in the long-term?

Faithfully answering these six questions will produce results more focused on the full set of factors driving uncertainty at the local scale where siting occurs than will the traditional BCA or EIS frameworks. A risk analysis approach for siting noxious facilities would be enhanced if practitioners incorporate three other processes and tools not widely use in traditional risk analysis but vital for siting (see Chapter 10):

1   More effective working group processes: Mandated public meetings will not go away, but we need to move toward setting up committees representing parties with a real stake in the outcome and allow committees and subcommittees to identify key issues and identify points of agreement and disagreement with a minimum of media-attention-grabbing histrionics (Chapter 7).

2   Deployment of analytical tools: In order to encourage productive interactions every opportunity should be taken to use analytical tools to identify,

examine, display, modify and debate the consequences of siting options (see Chapter 9).

3   Scanning: Systematically scanning for economic, social, political and technological trends is a way of forcing an organization and oneself to wrestle with comfortable and uncomfortable plausible events that will impact siting and non-siting options (Chapter 10).

Final thoughts

Traditional industrial location factors described in Chapter 2 remain – profit and cost minimization are core objectives. But how these play out in any country or specific area is not nearly as certain as it was when we believed that simple mathematical models or even the Varignon frame would identify the most economical location. The introduction of laws, regulations, improved environmental science and technology, the emergence of the challenges of public perceptions, values and environmental justice, and the assertion of government autonomy, especially local government, are exogenous factors that add layered levels of uncertainty to an investment in a noxious facility.

Arguably, the late twentieth and early twenty-first centuries challenge what many consider the excesses and arrogance of uncontrolled capitalism. This assertion is often the elephant in the room when siting analysts are engaged with others with different goals and values. I may be too sensitive, but I often feel a moral test is being administered to those of us who are analysts in the position of advising decision-makers or local populations. The augmented set of considerations discussed above and added in Figure 4.2 is seen by many as a long overdue effort to reestablish the hegemony of human and environmental health, cultural, social and other non-economic values in decision making. Whatever your view and role, conditions have changed and it is up to analysts and developers to respond with workable siting or non-siting options without compromising their personal integrity or their organization's credibility. Chapters 5 to 8, and 10 explore some of these options.

References

1   Krugman P. *Development, Geography, and Economic Theory*. Cambridge, MA, MIT Press, 1995.
2   McCann P. *The Economics of Industrial Location: A Logistics–Costs Approach*. New York, Springer, 1998.
3   Schemenner R. *Making Business Location Decisions*. Englewood Cliffs, NJ. Prentice-Hall, 1982.
4   Stafford H. *Principles of Industrial Facility Location*. Atlanta, GA, Conway Publishing, 1979.
5   Sule D. *Manufacturing Facilities: Location, Planning and Design*, 2nd ed. Boston, PWS Publishing, 1994.
6   Watts H. *Industrial Geography*. New York, John Wiley and Sons, 1987.
7   Rees J, Hewings G. Stafford H. *Industrial Location and Regional Systems*. New York, Praeger, 1981.

8   Barnes T, Gertler M, eds. *New Industrial Geography: Regions, Regulations, and Institutions.* Routledge Studies in the Modern World Economy, 22, New York, Routledge, 1999.
9   Chapman K, Walker D. *Industrial Location: Principles and Policies.* 2nd ed. Oxford, Blackwell, 1991.
10 Drezner Z, ed. *Facility Location: A Survey of Applications and Methods.* New York, Springer 1996.
11 Harington J, Warf B. *Industrial Location: Principles, Practice and Policy,* New York, Routledge 1995.
12 Hayter R. *The Dynamics of Industrial Location.* Chicester, Wiley, 1997.
13 Laulajainen R, Stafford H. *Corporate Geography: Business Location Principles and Cases.* Dordrecht, Kluwer Academic, 1995.
14 Perlstein R. *Nixonland: The Rise of a President and the Fracturing of* America. New York, Scribner, 2008.
15 Train R. "Prescription for a Planet: the Ninth Bronfman Lecture," in M. Greenberg, Voices from the past, Russell E. Train: A Leading Environmental Figure of the 1970s, *American Journal of Public Health,* 100, 4, 604–607, 2010.
16 Greenberg M, Goldstein B, Anderson E, Dourson M, Landis W, North DW. Whither Risk Assessment: New Challenges and Opportunities a Third of a Century After the Red Book. *Risk Analysis.* 35, 11, 1959–1968, 2015.
17 Greenberg M. *Explaining Risk Analysis: Protecting Health and Safety.* New York, Earthscan, 2016.
18 Jasanoff S, Nelkin D. Science, Technology, and the Limits of Judicial Competence. *Science,* 214, 1211–1215, 1981.
19 Greenberg M, Anderson R. *Hazardous Waste Sites: The Credibility Gap.* New Brunswick, NJ, Center for Urban Policy Research Press, 1984, republished Transaction Press, 2012.
20 Quarles J. Federal Regulation of New Industrial Plants, *Environment Reporter.* Monograph no. 28, 10, 1, May 4, 1979, 51pp.
21 Fortuna R. Ninety-Nine Ways to Lose and One Way to Win, Beating the Odds at Advancing the Proper Treatment of Hazardous Waste. Washington, DC. Conference: The Practical Politics of Hazardous Waste Management, Washington DC, Capital Hilton, May 5–6, 1987. Paper copy.
22 Stafford H. The Effects of Environmental Regulations on Industrial Location, report on National Science Foundation Grant no. SES_8024562, Cincinnati, June 1983, paper copy.
23 Slovic P, Fischhoff B, Lichtenstein S. Facts and Fears: Understanding Perceived Risk. 181–216 in Schwing R, Albers W, Jr. *Societal Risk Assessment: How Safe is Safe Enough?* New York, Plenum Press, 1980.
24 Easterling D. The Vulnerability of the Nevada Visitor Economy to a Repository at Yucca Mountain. *Risk Analysis.* 17, 5, 635–647, 1997.
25 Slovic P, Layman M, Kraus N, Flynn J, Chalmers, Gesell G. Perceived Risk, Stigma, and Potential Economic Impacts of a High-Level Nuclear Waste Repository in Nevada. *Risk Analysis.* 11, 683–696, 1991.
26 Sjoberg L. Local Acceptance of a High-Level Nuclear Waste Repository. *Risk Analysis.* 24, 3, 737–749, 2004.
27 McClelland G, Schulze W, Hurd B. The Effect of Risk Beliefs on Property Values. *Risk Analysis.* 10, 4, 485–497, 1990.
28 Nelson J. Three Mile Island and Residential Property Values: Empirical Analysis and Policy Implications. *Land Economics.* 57, 363–372, 1981.
29 Jenkins-Smith H. Reversing Nuclear Opposition: Evolving Public Acceptance of Permanent Nuclear Waste Disposal Facility. *Risk Analysis.* 31, 4, 629–644, 2011.
30 Slovic P, Finucane M, Peters E, MacGregor D. Risk as Analysis and Risk as Feelings: Some Thoughts About Affect, Reason, Risk, and Rationality. *Risk Analysis.* 24, 311–322, 2004.

31 Clark D, Nieves A. Comparison of Noxious Facilities' Impacts for Home Owners Versus Renters. Argonne National Laboratory. ANL/DIS/PP – 86421, 1993.

32 Sandman P. *Responding to Community Outrage*. Washington, DC, AIHA, 1993.

33 Lowrance W. *Of Acceptable Risk: Science and the Determination of Safety*. Loss Altos, CA, William Kaufmann, 1976.

34 Greenberg M. Energy Policy and Research: The Underappreciation of Trust. *Energy Research and Social Science*. 1, 152–160, 2014.

35 Public Data Access, Inc. *Toxic Wastes and Race in the United States*. New York, Commission for Racial Justice, United Church of Christ, 1987.

36 Bullard R, Mohai P, Saha R, Wright B. *Toxic Wastes and Race at Twenty, 1987–2007*, Cleveland, United Church of Christ, 2007.

37 U.S. GAO. *Siting of Hazardous Waste Landfills and the Correlation with Racial and Economic Status of Surrounding Communities*. Gaithersburg, MD, GAO, 1983.

38 Baden B, Coursey D. The Locality of Waste Sites Within the City of Chicago: A Demographic, Social, and Economic Analysis. *Resource and Energy Economics*. 24, 53–93, 2002.

39 Dunlap R, Kraft M, Rosa E. eds. Public *Reactions to Nuclear Waste: Citizens' Views of Repository Siting*. North Carolina, Duke University Press, 1993.

40 Zimmerman R. Social Equity and Environmental Risk. *Risk Analysis*. 13, 649–666, 1993.

41 Anderton D, Oakes J, Egan K. Environmental Equity: The Demographics of Dumping. *Demography*. 31, 229–248, 1994.

42 Greenberg M. Proving Environmental Inequity in Siting Locally Unwanted Land Uses. *Risk: Issues in Health & Safety*. 4, 235–252, 1993.

43 Bullard R. *Dumping on Dixie*. Boulder, CO, Westview Press, 1990.

44 Edelstein M. *Contaminated Communities: The Social and Psychological Impacts of Residential Toxic Exposure*. Boulder, CO, Westview Press, 1988.

45 Gerrard M. *Whose Backyard, Whose Risk*. Cambridge, MA, MIT Press, 1994.

46 Brown P, Mikkelsen E. *No Safe Place: Toxic Waste, Leukemia, and Community Action*. Berkeley, CA, University of California Press, 1990.

47 Erikson K. *Everything in Its Path: The Destruction of Community in the Buffalo Creek Flood*. New York, Simon and Schuster, 1977.

48 Lerner S. *Sacrifice Zones: the Front Lines of Toxic Chemical Exposure in the United States*. Cambridge, MA, MIT Press, 2010.

49 Smary E. Environmental Justice: More than "NIMBY" and "LULU". Michigan Chamber of Commerce, author's paper copy, 7–13, 1999.

50 Chess C, Purcell K. Public Participation and the Environment: Do We Know What Works? *Environmental Science and Technology*. 33, 16, 2685–2692, 1999.

51 Hance B, Chess C, Sandman P. *Industry Risk Communication Manual*. Boca Raton, FL, Lewis Publishers, 1990.

52 Baretta S, Bozzolan S. A Framework for the Analysis of Firm Risk Communication. *The International Journal of Accounting*. 39, 265–2188, 2004.

53 McComas K. Public Meetings about Local Waste Management Problems: Comparing Participants to Nonparticipants. *Environmental Management*. 27, 1, 135–147, 2001.

54 Lake R. Negotiating Local Autonomy. *Political Geography*. 13, 5, 423–442, 1994.

55 Chryssolouris G, Mavrikios D, Papakostas N, Mourtzis D, Michalos G, Georgoulias K. Digital Manufacturing: History, Perspectives, and Outlook. *Proceedings Institution of Mechanical Engineers*. 213, B, 454–462, 2009.

56 Zhou Z, Xie S, Chen D. Future Development of Digital Manufacturing Science. Foundations of Digital Manufacturing Science. London, UK, 337–363, 2012. http://link.springercom/chapter/10.1007/978–0-85729–564–4-9#page. Accessed April 5, 2017.

57 Hartmann B, King W, Narayanan S. Digital Manufacturing: the Revolution Will Be Virtual. August 2015. www.mckinsey.com/business-functions/operations/our-insights/digital-manufacturing-the-revolution-will-be-virtualized. Accessed March 27, 2018.

58 Hessman T. 2017. Brilliant: How GE Is Leading Manufacturing's Digital Revolution. www.industryweek.com/industryweek-manufacturing-technology-conference-expo/brilliant-how-ge-leading-manufacturings-digita. Accessed March 27, 2018.

59 Finch H. 2016. Chemical Industry More Open to Digital Innovation. www.icis.com/resources/news/2016/11/10/10053266/insight-chemical-industry-more-open-to-digital-innovation. Accessed March 27, 2018.

60 McCue T. Digital Manufacturing Path to Future with Proto Labs. *Forbes*, May 23, 2016. www.forbes.com/sites/tjmccue/2016/05/23/digital-manufacturing-path-to-future-with-proto-labs/#7ee69dcd1cca. Accessed March 27, 2018.

61 Johnson E. *Sustainability in the Chemical Industry*. New York, Springer Publishers, 2012.

62 Sharma S, Henriques I. Stakeholder Influences on Sustainability Practices in the Canadian Forest Products Industry. *Strategic Management*. 26, 2, 159–180, 2005.

63 GreenBlue Launches Environmental Paper Assessment Tool (EPAT) 3.0. Greenblue. org/press/greenblue-launches-environmental-paper-assessment-gool.et_3–0/ Assessed April 5, 2017.

64 Greenberg M, Weiner M, Mayer H, Kosson D, Power C. Sustainability as a Priority at Major U.S. Department of Energy's Defense Sites: Surrounding Population Views. *Sustainability*. 6, 2013–2030, 2014.

65 Sustainability Performance Office. USDOE. Plans and Reports. 2016. https://energy.gov/management/spo/plans-and-reports. Accessed March 27, 2017.

66 Dunlap R, Mertig A. The Evolution of the U.S. Environmental Movement from 1970 to 1990. *Society and Natural Resources*, 4, 209–218, 1991.

67 Greenberg M. *The Environmental Impact Statement After Two Generations: Managing Environmental Power*. New York, Routledge, 2011.

5   Becoming less noxious

Introduction

This chapter has three objectives:

- Define pollution prevention actions, including substitution of raw materials, recycling, energy recovery, and treatment.
- Define continuous safety improvements, including inherently safer designs, cyber security, and resilience.
- Illustrate their use with current efforts of Merck & Co, including its non-U.S. and Canada operations.

Becoming less noxious

In 1974, I attended a meeting about water and air pollution in Philadelphia. A government official said that he envisioned a day when chemical, paper and metal plants would have no emissions. The group, including me, did not believe that stage would be reached any time in the near future. Almost a half century later, we have not reached it. However, pollution prevention, inherently safer designs, resilience plans and security practices have reduced the chances of acute events and chronic exposures. The advantages and disadvantages of a generic less noxious policy are summarized in Table 5.1.

Pollution control was the major environmental agenda of the 1970s and 1980s, as U.S. laws and regulations described in Chapter 4 sought to reduce emissions and keep them out of the air and water. After concentrating on treating wastes before emission or shifting from air and water emission to land-based solutions, companies began to shift gears away from end-of-pipe choices to preventing chronic emissions, and reducing the risk of acute events. The major focus initially was pollution prevention (P2), and now increasingly includes inherently safer designs (ISD), enhanced security, and resilience.

Table 5.1 Advantages and disadvantages of a less noxious facility policy

Criteria	Advantage	Disadvantage
Raw materials and energy	Less raw materials needed and less stored on site that can cause hazard events	Changes could undermine some local resource-based economies
Labor	Higher paid jobs to design and manage complex at least partly automated systems	Loss of some semi-skilled jobs, especially to automation
Markets	Advantage to serve customers that support safer and less noxious companies	Markets shift and not every company will install best practices in every market
Capital	Save in the long run by requiring less resources and avoiding legal challenges, operation slowdowns and shutdowns	Capital investments needed for improvements, monitoring and operations
Land	Less land needed for storage and operations. Activities might be less intrusive	On-site construction to upgrade systems may impact surrounding areas and be considered unfavorably by local officials, publics, and other developers, at least in the short run
Local cooperation	Advantage if there is a history of cooperation and organization can present a strong proactive position that addresses local government concerns	Disadvantage if there is history of hostility, and if the new site or expansion heighten concerns about the credibility of the applicant
Infrastructure, including transportation	Major savings if less is needed and if the new systems are more readily monitored	Problem if the existing infrastructure are outdated and need to be entirely replaced, which would cause issues in the short run
Agglomeration	Advantage if there are benefits of concentrating production at a few sites for recycling, adaptability, security and other reasons	Even with pollution prevention and continuous safety improvements, local population may be distressed by the amount of hazardous material in a few or a single location
Environmental science and technology	Reduced potential for bioaccumulation of toxins, corrosivity, flammability, reactivity, radioactivity, toxicity, etc. Fewer workers exposed to hazards	Possible to create a new hazard by reformulation, product substitution, and recycling at the same or different site
Public concerns and environmental justice	A less hazardous facility helps mute concerns, at least among some residents	Some will see concentrating at fewer locations, even if they are safer, more evidence of environmental and social injustice
Local government autonomy	Influential local government officials may be more likely to support a less hazardous facility	Some local officials may oppose a facility irrespective of its improved safety level

Pollution prevention

During the 1990s, pollution prevention was part of the U.S. government effort to replace so-called "command and control" regulations with options that allowed companies to meet required targets in their own way, in other words, the "cookbook" of mandated processes and technologies was to be replaced by individually designed solutions.[1-3]

State pollution prevention laws appeared in the 1980s and by the mid-1990s, every state had at least one program to help industries reduce waste. The Office of Technology Assessment's 1986 report[1] helped persuade federal and state officials that P2 was a good policy. The Pollution Prevention Act of 1990 (P.L.101–508) offered a hierarchy of approaches, with source reduction most preferred and disposal least preferred:

- Source reduction;
- Recycling;
- Energy recovery;
- Treatment;
- Disposal.

Source reduction practices include but are not limited to the following:[1-5]

- Reformulating products;
- Modifying equipment, processes and technology;
- Substituting raw materials;
- Improving inventory control, maintenance, and training.

The list of specific actions that fall under these four is broad, for example, audits, waste and fugitive emission inventories, housecleaning, reduction of materials accumulated in storage tanks, leak detection monitoring, and many others.

Recycling is second on the OTA's list of P2 options and widely used when the waste is valuable and close to a potential user. So, for example, solvent recycling has been common for many years. On the other hand, wood was so ubiquitous that recycling and reuse to produce energy required the value of raw wood and the cost of disposing of it to increase before wood waste became a valuable raw material.

The EPA established an Office of Pollution Prevention in 1988 and developed a P2 strategy, including to incentivize states. By the mid-1990s, during the early years of the P2 program, case studies documented some of the success stories by exemplary companies. However, as the programs evolved it became obvious that companies, in essence, were making relatively easy and inexpensive changes, so-called "low hanging fruit," although their changes did reduce pollution.[6-9]

For example, The Business Roundtable conducted a benchmarking study published in 1993[9] based on six industry leaders in P2: Dupont, Intel, Martin

Marietta, Monsanto, Proctor & Gamble, and 3M. The report had multiple findings, and I have paraphrased several key observations from that report:

- All of these had strong upper management support for P2, including making it part of management goals and objectives;
- Management made sure that key leaders from different departments worked together;
- Facilities used the ingenuity of staff to determine the best methods;
- Facilities reported progress towards goals and objectives on a monthly or quarterly basis to make sure that everyone realized what was happening;
- The changes had to be cost effective and be part of the standard capital request process, which is notably different from environmental compliance projects;
- Every facility routinely used recycling and some used the entire P2 hierarchy;
- Even among these industrial leaders, more was spent on compliance than on P2 at these facilities, and participants wanted to move more toward P2 and away from compliance.

The TRI data (Chapter 1) has been used to assess the success of pollution prevention across tens of thousands of sites. Recognizing that not all emitters are required to report to TRI, a 1994 report by the U.S. General Accounting Office (GAO) observed that although many states had P2 programs, many did had not press for source reduction and instead supported recycling, treatment, and disposal. Individual facility reports sometimes attributed emission reductions to better emission measurement procedures and to focusing on easy to change operations. Some of the more thorough studies concluded that changing to P2 was hindered by institutional, financial, and technical barriers that were not always possible to quickly change.[10–12] Among the more interesting of these barriers was internal opposition by staff that had neither the expertise nor willingness to change manufacturing.

New Jersey had among the most demanding P2 programs.[13] About 700 industrial facilities were required to produce pollution prevention plans, summaries and progress reports. The state's Department of Environmental Protection randomly chose 6 percent of sites for review. Some interesting data emerged from this admittedly initial review.

- Prior to P2 only 24% reported tracking raw material use per unit of product (e.g., efficiency of chemical use); and only 14% tracked losses during production;
- The average number of days to prepare the plan was 13.5;
- The average company achieved 69% net cost savings as a result of implementing the plan;
- Crude benefit/cost ratios showed $5–$8 saved for every dollar spent;
- 74% of those studied concluded that the planning process was worthwhile because it led to savings, was preferred to regulations, led to better business

decisions, provided more accurate data, and led to the production and environmental management staff to work more cooperatively;

- Obstacles included that some required permit modifications, but that the biggest obstacle was that some staff resisted changing their practices.

Pollution prevention may not be what some had anticipated it would be. It has not ended emissions. However, as a reader and researcher about this subject for some years, my conclusion is that one approach did not fit all producers, primarily because factors discussed in Chapters 1 and 4 vary by facility. More specifically, I offer four observations about pollution prevention:

- Every form of pollution prevention seems to have worked for every producer in some places at some times;
- The variation within the same industry by facility is remarkable in terms of alternatives, information, employee cooperation or resistance, knowledge, customer preferences, environmental regulations, history of enforcement actions, and local community preferences;
- Some producers have been ethically committed to the idea but their bottom line profit motive resurfaces at critical decision points;
- Unless you are intimately familiar with the entire business, knowing which companies will use which method(s), at which time, and at which facility requires a crystal ball that does not exist.

How successful has the P2 program been? The initial case studies doubtless were overly optimistic, almost always focusing on success stories. In 2009, the Office of Inspector General (OIG) of the EPA, examined the database about P2 results.[14,15] On the one hand, among all of EPA's programs, the OIG rated the P2 program third. On the other hand, that relatively high rating needs to be seen in the context of three concerns reported by the OIG and widely discussed at meetings that I have attended. The EPA's metrics of success are based on emission reductions, not on impact of those emission reductions on human health. The database was submitted by cooperating organizations and these were not verified in the field. The program has been slow to follow up on previous assessments. In reality, I believe that EPA's part has been notably underfunded and lacks the capability to aggressively lead and respond.

It is hard to declare P2 a marked success, nor however, should it be declared a failure. An EPA report on production waste managed during the years 2003–2014 based on TRI reports shows why my assessment is equivocal.[16] It is hard to see much of change during that time period. Production waste managed by TRI facilities declined 4%. Disposal and other releases decreased by 14%, and treatment decreased by 7%. Recycling and energy recovery were about the same in 2003 and 2014. What is most interesting about the data is that there clearly was a decrease in emissions from 2003 through 2009. Then, the U.S. economy began to recover from the recession, and the emissions and waste management started increasing again. In other words, as value added by the industry sector went up waste managed and emissions followed.

These results are obviously disappointing to those that expected this program to be a panacea. But I do not believe that it has failed. Other EPA tables show that most of the changes related to P2 were made since the end of the recession are in better operating practices and raw material modifications, which are good signs. The reasons for lack of source reduction reported by EPA are entirely predictable. Specifically, three reasons account for almost 80% of the reported barriers to source reduction:

- 41% – no known substitutes or alternative technologies;
- 20% – pollution prevention was adopted, and additional reduction is not technically or economically feasible;
- 18% – customer concern about a change in the product.

While these issues continue to confound P2 advocates, several government-driven steps have pushed it along. One is California's Proposition 65, which pressures industry to reduce the use of toxins because the information is disclosed to the public. Second, the EU's Registration, Evaluation, Authorisation, And Restriction of Chemicals (REACH) is much more demanding than the U.S.'s TSCA legislation insofar as REACH requires new chemicals to be registered and evaluated before use, not after it, is already in the market. I believe that there will be another breakthrough in P2, but it is likely to follow from digital manufacturing. That change, as noted above, will not benefit those who may lose their job.

Can P2 be as successful as possible without government mandated changes? P2 is a sound investment for organizations with a high profile consumer products. But does it hinder creativity and limit flexibility? The author attended a conference at which presenters argued that pollution prevention options should be viewed through the lenses of optimization, and in fact they used a linear programming (Chapter 9) to optimize economic benefits and minimize costs among options such as,

- Reformulating products;
- Modifying equipment;
- Process and technology changes;
- Raw material substitutions;
- Improvement in inventory control;
- Maintenance and training;
- Recycling; and
- Treatment and disposal

We can also view P2 is a marker of business values and practices because it directs pressure at upper management to visibly support the idea, incorporate it into business plans, job responsibilities, and performance evaluations.[17] P2 also requires cross disciplinary corporate teams and up to date monitoring for it to succeed, which breaks the old pattern of corporate silos. With respect to government, the challenge is to move from an enforcement mentality to working with industry, including its own departments that are charged with using P2.

Continuous safety improvements

Three major interrelated policies make noxious facilities less hazardous and should make them less objectionable: inherently safer designs; cyber security; and resilience.

Inherently safer designs

On June 1, 1974, an explosion at a chemical plant near Fixborough in the United Kingdom killed 28 and injured 36, or almost everyone that was in the plant on that weekend day. Fortunately, the accident occurred on a weekend, or the toll would have been much higher. The outcry that followed led engineer Trevor Kletz to focus attention on inherently safer design (ISD). In essence, ISD is a logical way to systematically reduce expectable high-risk technological and human accidents that Perrow wrote about in 1984.[18] Kletz[19] offered recommendations that I have paraphrased:

- Use less hazardous raw materials in processes.
- Store as little hazardous material as possible on site.
- Design processes that use hazardous materials in safer ways and processes simpler to understand and use.
- Design facilities so that accidents and failures are prevented from cascading.

Kletz's ideas have evolved into a set of processes and policies. For example, Cozzani et al.[20] focused on reducing the likelihood of preventing cascading hazard events by conducting the risk analyses of the following:

- Primary accident scenarios;
- Scenarios that could propagate risk to other parts of the facility;
- Scenarios that involve secondary events triggered by the first one;
- Scenarios that propagate the event beyond the confines of the original event.

A 2002 survey of practitioners in 11 countries by Gupta and Edwards[21,22] found much less adoption if ISD than they had anticipated. The authors attributed the results to five factors:

- Lack of economic and benefit research;
- Lack of methods that demonstrate how to adopt the practices;

- Lack of desire to change current practices;
- Lack of knowledge among experts;
- Lack of enforcement by regulatory agencies.

Deadly events continued at several U.S. refineries and other plants, but no major actions were taken until an explosion occurred at a fertilizer plant near Waco, Texas on April 18, 2013. Following this event, President Obama appointed a U.S. government interagency task force headed by EPA and including the occupational safety and health administration (OSHA) and the U.S. Department of Homeland Security (DHS) to prepare a report and recommendations.

In 2016, the EPA called for some companies, including refineries and chemical companies, to incorporate ISD into their five-year plans.[23] Furthermore, companies that have had chemical accidents would be required to have an independent audit of their safety processes, and those that have had near misses would have root-cause analyses.

What has followed will be familiar to many readers – the often-practiced debate between rigid adherence to government regulations and flexible organizational accommodation to final goals without process specification. For example, the American Chemical Society[24] called for member companies to conduct vulnerability assessments, and they supported ISD and other approaches to increasing safety. But they noted that there is no magic formula for each facility and were concerned that a formulistic adoption of ISD might lead to inferior products, and new hazards, including environmental impacts that cannot be readily anticipated.

Some see this response as the industry dragging its collective feet, or it can be seen as a product of history in which companies were required to adopt specific technologies and processes, which sometimes were a bad fit for them. I do see a growing consensus to take the following steps in the chemical, petroleum, energy and many other industries; all are at the core of risk assessment and management practice:

- Identify hazards;
- Understand the processes that create hazards;
- Avoid these hazards and processes, if possible;
- Determine methods to reduce the severity of hazard events;
- Reduce likelihood of hazard events;
- Segregate hazardous processes to prevent acute events and reduce their consequences;
- Apply passive safeguards;
- Apply active safeguards;
- Apply procedural standards;
- Apply residual risk reduction measures.

Which of these to apply, when and for what period of time is the issue. Producers would like to avoid being held to a rigid set of actions on a rigid schedule, but some will wait as long as they can to reduce risk. Based on my experience, mixed with my optimistic personality, I believe that industry leaders are aware of the need to adopt ISD and will do so, thereby forcing more recalcitrant organizations to follow.

Cyber and physical security

Spies, saboteurs and other bad-intentioned people attacking facilities is the subject of many science fiction movies and books, and a reality during hot and cold wars. Data from the different sources show that globally utility power stations and petroleum sites are about half of the targets. But there is some data on every kind of noxious facility listed in Chapter 1. For example, U.S. nuclear power plants are deliberately surrounded by an area that is not accessible to the public. But there is always the possibility of an attack. Hence, they periodically have force-on-force tests to look for vulnerability, and the operators must take other measures to increase physical security.[25]

The U.S. Department of Homeland Security, as well as the EPA and the Department of Labor have been charged with physical security of chemical and allied facilities.[26] Threats typically come from physical intrusions and insider attacks. But increasingly cyber threats are a concern, including cyber espionage, crime ware, and web app attacks. The impacts include plant shutdowns, the theft of intellectual and personal information. In the case of noxious facilities threats include hazard materials emissions, overpressure of equipment and then ruptures, and exposures beyond plant boundaries. The idea of living near a chemical, petroleum or metal plant that can be disrupted is a local as well as a corporate and government concern and influences siting.

Cyber security is relatively new. Like any new threat, I assume, without any data to support my assumption, that people are more frightened about it than they are about physical intrusion, which they have seen movies about. The chemical industry, which is heavily automated and increasingly digital, has expressed considerable concern. For obvious reasons, data are hard to come by, but what is available shows more attacks. For much of this century, there was a general belief that hacking was motivated by people trying to make money, the idea was that hackers were interested in stealing personal data, so the threat was rated as an increasingly expensive inconvenience. Now, there is more concern that facilities and operations will be targets. Should these events become more common it is easy to imagine a local community strongly opposing a new facility or expansion of an existing facility, even if it creates jobs, pay taxes, and has an outstanding P2 program.

The cyber concern is part of overall security and extends to many noxious facilities addressed by the following programs:

- The North American Electric Reliability Corporation (NERC), which is certified by the Federal Energy Regulatory Commission [Critical Infrastructure Protection (CIP) cyber security reliability standards];[27]

- Executive Order 13650 [improve chemical facility safety and security with owners, operators and government];
- Chemical Facility Anti-Terrorism Standards (CFATS) which regulate chemical industries that "present high levels of security risk." (Section 550 of the DHS Appropriations Act of 2007). [Chemicals, energy, food, paints and coatings, explosives, electronics, and plastics included];[28]
- Homeland Security Presidential Directive -7 (HSPD-7) [focuses on both security and resilience to cyber threats as part of an effort to identify and prioritize critical resources and infrastructure].[29,30]

The reach of these requirements is wide and beyond the scope of this section to consider. I have listed several brief articles and strongly suggest typing in the U.S. Department of Homeland Security and then critical infrastructure sectors. These include the following that fit into the noxious industry definition, which includes almost all the sectors defined as noxious in Chapter 1:

- Chemical;
- Critical manufacturing (includes primary metals, machinery, electrical equipment, and transportation equipment);
- Defense industrial base;
- Energy;
- Nuclear reactors, materials, and waste.

Depending upon the specific sector, there will be a definition of what is included, a sector specific plan, resources, and in some cases training and contacts.

Resilience

When we are injured or ill, we have to try to return to normal or change to a new normal. Resilience can be challenging for a company, government organization and an area. For instance, after the Fukushima and Chernobyl nuclear accidents, a new normal had to be created for many of the people, the organizations, and impacted areas of Japan.

The impacts of major hazard events can disrupt areas outside the directly impacted zones. The failures at Fukushima jeopardized the Japanese nuclear industry, led other governments like Germany to develop a plan to eliminate nuclear power, as well as made many commercial nuclear companies and governments review their policy for certain reactor types and for the storage of spent nuclear fuel.

Resilience, in short, has become a much discussed topic, similar to sustainability, and it is practiced both through engineering/design and human performance. To gauge resilience, analysts must conduct a risk analyses to make the following determinations:

- Potential threats;
- Symptoms that a failure is imminent;

- Knowledge that a serious event is occurring;
- Protocols to be activated to reduce the risk when an abnormal event is occurring.

In this regard, one of the interesting outcomes from the Fukushima events was the role of local radio in communicating information,[31] a role that has not been well performed during many previous hazard events around the world.

Nearly every noxious industry has been responsive to calls for resilience and conducted risk assessments and management studies, built physical barriers to control events, has safety nets, and instituted a safety culture. While equipment failures are often identified as a culprit in hazard events, final reports typically show that human failures were the direct cause or could have reduced the impacts. For example, human failures are often associated with stress in the workplace, and outside of it; with a lack of training on how to respond to events; and a failure to encourage employees to use their ingenuity, creativity, adaptability, as well as their training when facing a stressful situation. In other words, physical adjustments through ISD and security scanning are necessary but insufficient. Eventually, noxious facilities will confront a problem that existing engineered systems cannot automatically manage.

The petroleum industry[32] has experienced more than its share of problems, and it along with others have been focusing on equipment performance, but on human failures, such as

- Physical and psychological factors that lead some people to perform more effectively under stress than others;
- Variations in ingenuity, creativity, adaptability under stress;
- Impacts of education and training on more effective responses;
- Inside and outside workplace conditions that influence responses to events.

Case study: Merck worldwide

Merck and Company known at Merck Sharp & Dohme outside the United States and Canada (hereafter called Merck) is the seventh largest pharmaceutical company in the world dating back to the Merck family and seventeenth century Germany (manufactured morphine). In the United States, Merck was founded in 1891 as a retailing establishment in New York City.[33] Merck has purchased other companies, been absorbed and merged more than a dozen times, with the most recent mergers changing not only its business but also its site needs.[34] Merck products have included the following:

- Streptomycin (anti-bacterial agent);
- Diuril (hypertension control);
- Gardasil (control of papilloma virus);
- Singular and Claritin (allergy control);
- Propecia/proscar (male urinary symptoms and baldness);

- Januvia and Primaxin (type 2 diabetes control);
- Vioxx (arthritis control, withdrawn after charges that it elevated risk of heart attacks and strokes);
- Coppertone (sunscreen).

In addition to these products, some readers will have seen and used *The Merck Index*, which was a list of chemicals and chemical compounds. My copy, the 11th edition,[35] which I still use, is about three inches thick and sits about two feet away from my desk. Merck no longer produces it.

In 2016 Merck reported revenues of almost \$40 billion (U.S. dollars) and net income of \$5.7 billion.[36] However, tens of thousands of people have left the company because of retrenchments, which have involved multiple facilities openings and closings (see below).

I chose Merck for this chapter for three reasons. First, I have visited the Rahway plant on various occasions; worked on a pollution prevention project focused on the Merck former headquarters in Rahway, New Jersey; have been on a committee with a former Merck CEO, and two of my former PhD students work for Merck as do two friends who have retired from the company. Second, we have conducted fieldwork in Rahway, which included asking residents about neighborhood quality. Third, Merck epitomizes the complex decision-making considerations that large globalized producers consider when they make siting decisions about pollution prevention. Part of this case study will focus on Merck as a large international company and part will focus on its U.S. operations. I believe that the reader will find portions of this case study resemble the industrial legacy cases presented in Chapter 3 and part of it will read like the image of an aggressive corporation looking to survive and prosper in a complex environment.

The case study focuses on pollution prevention and location and relocation issues. It has little to offer in regard to inherently safer designs, cyber security, and resilience because these data are not and should not be public because of security concerns.

A start in Rahway

In 2003, Merck and its host communities Rahway and Linden received a "smart growth leadership" award from New Jersey Future, a not-for-profit that promotes a high quality living environment through cooperative planning.[37] In 2003, Merck had been in Rahway for a century, where it had purchased 150 acres at a site about 20 miles southwest of its New York City. In 1926, the company moved its corporate headquarters to Rahway. Merck was praised by the foundation for building facilities and a design that resembled a college campus. While the northern part of the site had tanks and pipes and other obvious manufacturing paraphernalia, the southern part with its dark brick buildings resembles a typical college campus, with brick buildings, curved roads, trees and parking garages (Figure 5.1).

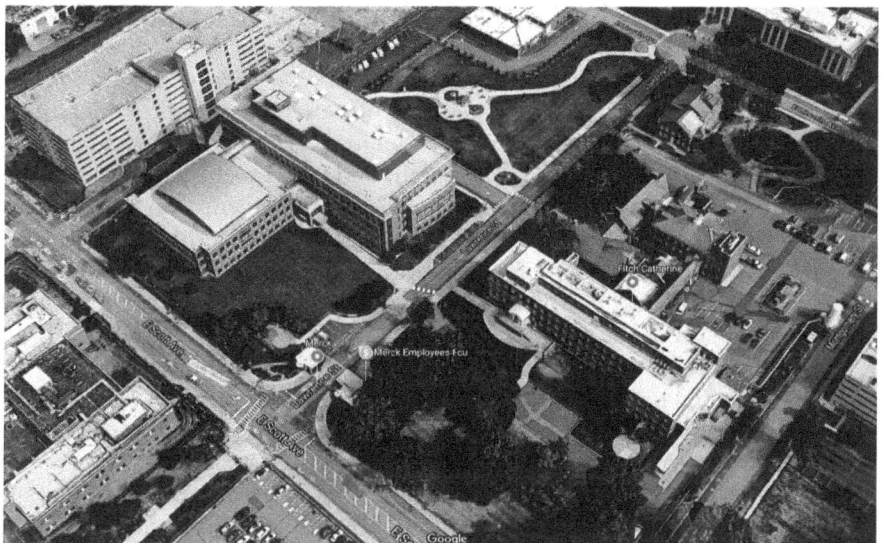

Figure 5.1 Merck Rahway site.

In 1992, Merck relocated its corporate headquarters to western New Jersey, ostensibly because there was insufficient space in Rahway to accommodate planned increase in research and development work. The fact that some senior managers lived in the town where the new corporate headquarters was located is a second reason for the relocation (see Chapter 4 for discussion of personal factors in site location), and the third is that the new location was halfway between Rahway and another major Merck factory.

Back in Rahway, the original site had 150 buildings and 4700 employees. Merck helped persuade New Jersey Transit to make Rahway a regular stop on its commuter line between New York City and Trenton, NJ, to the south, and Merck worked on a plan to expand Rahway operations. Merck, unlike some of the companies in Chapter 3, was not acting like it was ready to abandon its historic site with a contaminated legacy.

Furthermore, the 2003 New Jersey Future award was not the first time that Merck had been publicly praised, even by historically critical media. For example in 1992, Peterson,[38] writing for the *New York Times*, noted that the headquarters relocation did not "irk" its neighbors. Merck, the story noted, had informed the city well in advance that this was happening and indicated that research jobs were to be increased in Rahway. The article mentioned that Merck had been judged "America's most respected corporation" for six years by *Fortune Magazine*. Merck also committed to building a new research and development plant on the site, and a new day-care center for its employees. The new facilities would add $20 million in taxable property to Rahway's property. And the company donated money to the town for redevelopment planning. The NJ Future award represented a century of effort by the company to make itself indispensable to its workers, to listen to the surrounding jurisdictions, and to respond to problems that it had created.

For instance, in 2005 the EPA's regional office reported that the Merck site had underground water contamination, including pesticides and polychlorinated biphenyls from its manufacturing, research, and waste management activities. The EPA and the State of New Jersey agreed to a settlement plan and calendar for remediation. Merck has had similar agreements about other contaminated sites in New Jersey. In 2006, the company agreed to pay a $2.4 million fine for ground water pollution and three other New Jersey sites and as part of that agreement donated 10 acres of land for a restoration project.[39,40]

In the late 1980s when the first TRI data were made public, I learned that Merck had among the highest emissions of any facility in the state. A year later, the TRI data showed that emissions were 40% less and by 1998, 5/6th of the emissions had been eliminated. Merck, unlike many of companies, described in Chapter 3 and many others in New Jersey seemed to have lived a charmed life in the eyes of elected officials and the media. Yet, arguably, the praise they received was merited.

But what about the people who live near the site? In regard to demographics, EPA's EJ screen tool shows that the population within a mile of the Merck site is over 70% Afro- and Latino-American compared to about 55% for the city as a whole, and income of the surrounding area is about 15% lower. With that as context, in 1994, we conducted a survey of Rahway residents to assess the impact of the relocation of headquarters functions out of Rahway and to compare it with other developments in Rahway, such as construction of an incinerator.[41] The 360 responses focused on questions modeled on questions from the American Housing Survey. After correcting the sample responses for differences in gender, homeowners vs. renters, and education levels, the results were categorized by proportion of respondents who wanted to leave as a result of 21 different factors. Ten percent or more said that the following six caused them to want to leave the area:

- Trash incinerator – 28%;
- Airplane or train noise – 22% (there is a small plane airport in the area);
- Odors or smoke – 19% (identified by respondents with the incinerator and truck traffic);
- Motor vehicle noise, heavy traffic – 17% (heavily used Route 1 cuts through the center of the city);
- Oil refinery and tank farm – 15% (found in the town);
- Chemical plant, other manufacturing – 10% (Merck was among this group).

Follow-up one on one conversations with residents found that indeed the new incinerator was extremely bothersome.[42] Not only did they consider it an

Table 5.2 Neighborhood quality ratings of Rahway, 1994

Place and year	Excellent, %	Good, %	Fair, %	Poor, %
Rahway, 1994	6.7	52.7	35.3	5.3
U.S., 1991	33.6	52.6	11.2	2.6
Northern N.J., 1991	32.9	54.7	9.7	2.7
Marcus Hook and southern Chester, PA, 1993	5.6	37.1	40.6	16.8
East Elizabeth, NJ,1993	4.1	34.8	43.7	17.4

Source: Data from multiple tables and discussions.[42,43]

eyesore, but trash sometimes fell on streets, lawns, and garbage trucks used local streets instead of highways. In these verbal discussions, Merck was seen as company that had not abandoned the city.

Table 5.2 compares the results of this survey in regard to neighborhood quality ratings with other neighborhood taken at roughly the same time.[43] Rahway's ratings are much lower than the national ones and those of the northern part of the state. Nationally and within the northern part of the state, one-third rated their neighborhood as excellent and more than half rated it as good quality. Only 13%–15% rated their neighborhood as fair or poor quality. This contrasts with the two blighted and crime-impacted impacted areas in Chester and Marcus Hook, Pennsylvania located just south of Philadelphia and with eastern portion of the city of Elizabeth, New Jersey where close to 60% rated their neighborhoods as fair or poor. Rahway is closer to these last two than the national and northern New Jersey profiles but had a larger share in the good and much smaller share in the poor quality category than last two.

In short, Rahway has changed from a small city with solid industrial base to one with many of the same issues as many former industrial cities in the United States. In this context, Merck is the lone industrial survivor and continues some operations in the city, rather than having abandoned it completely.

International presence, competition and image

While revolutions in science and engineering have changed how and where production occurs, large companies, like it or not, have had to assume a much more worldly proactive and progressive public stance about issues that impact their public image. In 1998, Berry and Rondinelli[44] documented this change in regard to environmental protection. Three center pieces of a proactive corporate policy are preventing pollution, making products available to poor customers at discounted prices, and providing improved working conditions. This is not necessarily motivated by altruism, but some of that is present. It has been motivated by evidence that being socially and environmentally responsible leads to better public images and higher stock prices.

High profit margin pharmaceutical companies, like Merck, are in a strong position to make these kinds of changes. In regard to environmental protection,

Schering-Plough merged with Merck in 2009, and Merck moved to the former Schering-Plough headquarters in Kenilworth, NJ. The company's 2009 report on safety, health and environment is 89 pages of statements centered on environmental concern, evidence of reductions in emissions and notes about environmental awards.[45] Paraphrasing, the first slide in the report from upper management says that the company:

- Commits to protecting health and safety of all employees;
- Sets a goal of zero accidents and injuries; and
- Complies with laws and regulations, and protects the environment in areas hosting plants.

If my college and graduate professors had read this report, they would have been shocked, and some I know would not have smiled because it would seem to them like a waste of resources. The report then documents continuous improvements in environmental protection. For example, the report presents a chart that shows that safety, health, and environmental capital investments rose from $10.6 million in 2006 to $34.1 million in 2008, and overall expenditures in operations related to safety, health and the environment rose from $38.5 million in 2006 to $84.1 million on 2009. The document lists how it builds corporate international policies, sets standards and how these move from corporate to individual facilities. One of the criticisms often heard during the previous era was that health and environment were centered in a single corporate department and isolated from upper management. This report lists ten vice presidents or directors that are charged with carrying out the company's environmental programs and planning process linked to three- to five-year planning and annual plans.

The 2009–2011 report lists what needs to be done to achieve what it calls the "highest standards:"

- Leadership and involvement;
- Clear responsibility and accountability;
- Commitment to zero incidents;
- Effective management systems;
- Integration of safety, health and environment throughout the business;
- Sustainable products and processes;
- Open communication;
- Full compliance.

The company then proceeds to summarize each of the above, pointing to specific documents that provide additional data. For example, it discusses the needs for audits and certifications and lists 14 certifications obtained by businesses in 11 countries between 2002 and 2009. While safety, health and the environment are the focus of the report, social concerns are noted. For example, the following social issue priorities for 2008–2009 are listed:

- Patient safety;
- Occupational health and safety/environment/security;
- Promote scientific expertise and innovation;
- Diversity;
- Corporate governance of environmental health and safety.

Later in the report, brief summaries of improvements in safety, health and the environment are noted by factory by location across the world. The impact on emission reductions, health and safety, and how much money was saved are both listed when that information was known.

Other lists and tables appear throughout. Two of these are notable because they follow upon each other. The first is a list of opportunities that I paraphrase below:

- Cost reduction;
- Reduce lead times and reduce process flows;
- Lower the risk of failing to meet compliance requirements;
- Meet or exceed customer expectations;
- Increase safety and environmental quality;
- Increase colleague commitment and moral.

For continuous improvement metrics they list reductions in:

- Injury and illness rates;
- Days away;
- Reportable environmental incidents;
- Regulatory inspections with actions needed;
- Annual utility, water, hazardous waste, wastewater discharge and sales fleet.

The unambiguous message is that a cleaner and safer operation means less cost. When Merck and Schering-Plough merged, the stated corporate concern about environment and social responsibility did not change. For example, the 2015/2016 corporate responsibility report includes a discussion of how the company was working with health care providers to invest in unmet health needs. In 2011, Merck was investing in 53% of the 20 major global burdens of illness, and in 2014 it was 88%. Merck reduced prices of drugs in 114 poor countries. In 2013, Merck was honored for the 26th time for being among the 100 companies committed to a progressive workplace environment, improving the pay, status and working conditions of its female employees, as well as providing child care, flexibility, time off and leave. With some additional research, I found that Merck had received 48 awards from late 2014 to 2016 in the broad categories of philanthropy, ethics, transparency, support of employees, a broad category called global recognition.[46,47]

Five of the 48 awards are environmental sustainability.[46,48] These awards and accomplishments go back decades. In 1990, I was at a meeting at which then

New Jersey Governor James Florio touted New Jersey's P2 program as a win-win for companies. During the speech, he used Merck as the example of company that was embracing P2. The fact that Governor Florio praised Merck was no small achievement. As a member of the U.S. Congress, Florio wrote the first drafts of superfund legislation (CERCLA). As a colleague of the former governor's for decades, I can unequivocally state that he did not quickly praise companies.

Fast forward a quarter of a century when Merck receives the 2016 Energy Star Sustained Excellence award from the EPA for its efforts to reduce energy use at various sites. It received the same award in 2017.[49] *Newsweek* ranked Merck number 150 out of 500 global companies in greening and number 82 in the U.S. Merck had also received prominent awards that directly relate to pollution prevention and ISD. In 2010, it received the Presidential Green Chemistry challenge award from the U.S. EPA for developing a biocatalyst would increase the yield of sitagliptin and reduce byproduct wastes.[50] (Sitagliptin treats people with type 2 diabetes).

In addition to awards, Merck reported that audits by regulatory agencies or clinical trial investigators did not lead to any significant fines, penalties, warning letter or product seizures during the period 2001–2014. In terms of product recalls, the proportion was less than 0.22%. In regard to environmental goals, worldwide goals for the period 2009–2015 included a 25% reduction in global water use, 15% reduction in greenhouse gas emissions, and a 30% in waste production. Each of these goals was met or exceeded.[46]

In 2007, the company reported on what it calls "success stories" in energy management.[48] It promised to reduce energy use 25% between 2000 and 2005 while increasing production and achieved the goal by 2004. Along with this decrease was a 21% saving in energy expenditure and 250,000 tons of avoided carbon emissions. In New Jersey, the company installed solar panels at the Rahway site, added energy efficient lighting and HVAC systems, and increased training about energy savings. In 2004, they set a goal reducing worldwide energy use by another 25%.

Companies rarely publicize their security, resilience and inherently safe design programs because of security concerns and to maintain competitive advantage. I found a published paper that illustrates the company's ISD approach, one that focused on efforts to deal with worker safety. The process chemistry department found that some reactions could be highly energetic and pose risk to lab scientists and others even at the bench scale.[51] They listed hazardous reactions that must be automatically evaluated, including for example, nitrations, peroxidations, dinitro or trinitro compounds, hydrazine and others. These were included because they have the potential for rapid energy release. If an experiment includes these elements, various levels of consultation are required, including discussion with the supervisor, filling out a paper report that includes a literature review, a chemistry assessment of what might happen, securing preliminary data, and obtaining approval from the Scientific Advisory Committee and Environmental and Process Safety Engineering designee before doing the experiment.

At this point in this case study, some readers doubtless believe that I own Merck stock. I do not, and as the discussion above demonstrated, the company has had environmental issues. It has not all been awards. When it has had problems, the company appears more than willing to address them.

New economic realities and opportunities

The other side of the progressive company with proactive objectives and success at achieving national and international awards is that the company at times has been economically stressed and has made major cuts.[52–56] For example, after the Merck and Schering-Plough merger, in 2011, the company tripled the number of manufacturing plants from about 30 to 90. It began cutting employees – about 36,000 over five years. Company management needed to reduce costs and streamline production. Cuts came from plants in Italy, France, Ireland and Puerto Rico, also from among its original site in Rahway and the Schering's site in Kenilworth, New Jersey.[52]

In July 2015, Baldwin[52] reported that lab research jobs were being cut at Rahway and Kenilworth, as well as at the Merck facility in North Wales, Pennsylvania. Then it became clear that it was talking about focusing on biomedical research and locating close to the places where the best biomedical scientists were clustered, which it identified as the Cambridge, Massachusetts, and San Francisco areas to focus drug discovery, preclinical studies and early development research.[55] This cut amounts to about 10% of the jobs in early-stage research and development work at those three locations.[53–55] Carroll[55] added that the cumulative layoffs began after the merger but were exacerbated by problems in new drug approval clinical failures.

My friends retired from Merck with good benefit programs, including continuing access to the on-site fitness center. But the company eliminated their research program.

Before ending this case study, I return to the first major part of Chapter 1, which compared world manufacturing growth in hot spots, including a comparison of the United States and China. In 1994, Merck opened a plant in Hangzhou, China, which when I visited was told that it was to be a major technology hub. With a metropolitan population of over 20 million and a location near hundreds of millions of people and along the high speed rail that links Beijing and Shanghai (about five hours on the high speed train). Over the decades, Merck sold 40 vaccines and medicines in China.

In 2011, Merck established its Asian R&D headquarters in Beijing and stated that it planned to invest $1.5 billion in R&D in China during the period 2011 through 2016. Merck's press release notes that it will pursue this large, rapidly growing, and aging population. In 2013, the company opened a new manufacturing facility in Hangzhou with a $120 million investment to add to the existing manufacturing plant, and a site for marketing in Shanghai employing more than 5000.[57] None of these changes and the locations for these sites should surprise readers of this book. In the United States, it would have been the equivalent of locating along the Eastern Seaboard between Boston and Washington, DC, with New York City, Philadelphia, Baltimore, and others in the middle. The new plant, in short, is located in the major economic and population center of China. It is rational for a major international pharmaceutical company to do precisely what Merck is doing in China.

How was this new plant designed? The company notes that the plant will produce medicines to manage cardiovascular diseases, diabetes, infectious diseases, and respiratory and bone diseases. In regard to the environment, the release says that the facility is "fully compliant with manufacturing facilities worldwide." The facility has engineered systems to manage air, waste and water emissions.[57] Overall, the non-U.S. and Canadian parts of Merck serve 140 countries from 72 factories. I have not been able to visit any of them, but I would expect to find that the company is adhering to its written assertions about health care and the environment as part of its effort to build its market share, which includes meeting health and environmental expectations.

Writing in 2003, Hawthorne[34] captures my view of Merck as a company that has used many of the methods in these case study chapters. The book *The Merck Druggernaut* has several fascinating descriptions. Early in the book, Hawthorne (p. xii) says that he considered calling the book "the last good drug company," but later he notes that he also thought about calling it "the struggles of a once-shining star" (p. 251). Merck's siting patterns and choices in the twenty-first century are in many ways similar to the kinds of choices documented by Gortler[33] in his history of the first 70 years of the company. It made decisions, some with painful consequences for some but major benefits for others. Some of the decisions made in New Jersey helped and then hurt several communities. But then compared to many others that had operations within 20 miles of Merck (Chapter 3), the pain to the local community was less severe and more gradual in regard to economic impacts and especially in regard to environmental consequences.

International examples

There are interesting examples of pollution prevention, substituting raw materials, and waste management from other countries. In Vietnam, water pollution has reached critical levels due to population growth, urbanization, and ever developing industrialization.[58] As one of the main drivers of Vietnam's growth, industry's share in GDP rose from 22.7% to 41.1% between 1990 and 2010. Currently there are more than 200 industrial zones in Vietnam and about one million cubic meters of untreated wastewater is generated daily.[59] In 2012, the World Bank approved the Industrial Pollution Management Project, which will take place in Vietnam's most industrialized provinces: Dong Nai, Ba Ria Vung Tau, Nam Dinh and Ha Nam. The World Bank funds the project through its International Development Association (IDA) and The Ministry of Planning and Investment of Vietnam will be responsible for implementation.[60] The project is expected to cost about $58 million for building eight Centralized

Effluent Treatment Plants (CETP), constructing monitoring infrastructure, and improving compliance with industrial wastewater treatment regulations.[60,61] The most recent report from the World Bank reveals that four Centralized Effluent Treatment Plants began operation as of March 2017 in order to manage industrial water pollution.[61] Additionally, the project involves reviewing and revising laws, legal documents, and regulations at central and provincial levels, which were recently criticized by economic and environmental specialists as insufficient since they can be dodged by international investors.[62]

In Europe, one of the largest polymer companies Covestro, formerly Bayer MaterialScience, has been working on replacing petroleum with CO_2 in the production of high-tech plastic materials. Project "Dream Resource" aims to substitute oil-based raw materials with CO_2 that would reduce energy input and carbon emissions, while transforming CO_2 into useful raw material.[63] Covestro collaborated with RWTH Aachen University for the project, which is funded by German Federal Ministry of Education and Research (BMBF). It was reported that feasible production of foam material is achieved with 20% fewer greenhouse gases emissions.[64] In 2016, the project won the European Chemical Industry Council's (Cefic) Responsible Care Award in the environment category and was recognized as one of the best practices among chemical companies.[65] Recently, the CEO of the company mentioned that they are considering licensing the technology to countries like China to encourage widespread operation.[66]

Unilever is an international consumer goods company, which has more than 600 sites in 70 countries. Its factories produced about 140,000 tons of waste in 2008. In the same year, the company committed to send zero non-hazardous waste to landfills by 2020.[67] In its factories, warehouses, distribution centers, and offices, Unilever implemented a "reducing, reusing, recovering or recycling" approach for different non-hazardous waste streams.[68] As a result, the company achieved this target in all of its sites in 2016.[69] Apart from that action, Unilever is also known for its sustainable initiatives on waste management, especially in packaging. In 2014, they collaborated with Innovate UK, a United Kingdom government-funded innovation agency, to focus on recyclable packaging.[70]

Final thoughts

Practicing pollution prevention, and incorporating inherently safer designs, improving security, especially cyber, and resilience into planning is increasingly common among major companies and government agencies who read twenty-first century tea leaves, that is, recognize that the vast majority of people have expectations about organizational behavior vis-à-vis their workers and their families, and their environmental health programs.

Companies and agencies in search of new locations want to avoid the fierce confrontations that often focused around environmental and social justice starting in the 1980s described in Chapters 3 and 4. In essence, companies like Merck and government agencies want to walk into a negotiation with a progressive record in regard to the environment and social justice. They want to be able to demonstrate that the organization is moving up the ladder toward more protection and justice not down it to the bottom. The challenge for organizations will be impacted by any local history, but even a conflict laden one is not insurmountable if the organization is trusted and can take steps to increase that trust.

References

1 Office of Technology Assessment, U.S. Congress. *Serious Reduction of Hazardous Waste for Pollution Prevention and Industrial Efficiency* (OTA-Ite-317). Washington DC, U.S. Government Printing Office, 1986.

2 Freeman H. *Industrial Pollution Prevention Handbook.* New York, McGraw-Hill, 1994.

3 Shen T. *Industrial Pollution Prevention.* New York, Springer 1995.

4 U.S. EPA. *EPA Pollution Prevention Accomplishments: 1994.* Office of the Administrator. Washington, DC. EPA 100-R-95-001, 1995.

5 U.S. EPA. *Facility Pollution Prevention Guide.* Washington, DC, Office of Research and Development. EPA/600/R92/088.

6 Dorfman M, Muir W, Miller C. *Environmental Dividends: Cutting More Chemical Wastes.* New York, Inform, Inc. 1992.

7 Sarokin D, Muir W, Miller C, Sperber S. *Cutting Chemical Wastes: What 29 Organic Chemical Plants Are Doing to Reduce Hazardous Wastes.* New York, Inform, Inc., 1985.

8 Bresnan J. An Analysis of Private, Public, and Non-Profit Industrial Pollution Prevention Incentives. *The Environmental Professional.* 16, 1, 57–65, 1994.

9 AT&T for the Business Roundtable. Facility Level Pollution Prevention Benchmarking Study. November 1993. Infohouse.P2ric.org/ref/28/27926.pdf. Accessed April 18, 2017.

10 Cebon P. Corporate Obstacles to Pollution Prevention. *EPA Journal.* 19, 3, 20–22, 1993.

11 Byers R. Regulatory Barriers to Pollution Prevention. *Pollution Prevention Review.* 2, 1, 11–19, 1991/1992.

12 Ashford N, Ayers C, Stone R. Using Regulation to Change the Market for Innovation. *Harvard Environmental Law Review.* 9, 419–466, 1985.

13 *Chemical & Engineering News.* Toxics Release Inventory Shows Pollution Decline. 1992, June 1, p. 16.

14 Department of Environmental Protection, State Of New Jersey. Early Findings of the Pollution Prevention Program. Trenton, NJ, Paper Copy, June 1995.

15 Office of Inspector General. U.S. EPA. Measuring and Reporting Performance Results for the Pollution Prevention Program Need Improvement. January 28, 2009. www.epa.gov/oig/reports/2009/20090128-09-p-0088.pdf. Accessed April 19, 2017.

16 U.S. Environmental Protection Agency. Tri National Analysis 2014: Pollution Prevention and Waste Management. www.epa.gov/trinationalanalysis/pollution-prevention-and-waste-management-2016-tri-national-analysis. Accessed March 27, 2018.

17 Tesh S. *Hidden Arguments.* New Brunswick, NJ, Rutgers University Press, 1988.

18 Perrow C. *Normal Accidents: Living with High-Risk Technologies.* New York, Basic Books, 1984.

19 Kletz T. *Process Plants: A Handbook for Inherently Safer Design.* Philadelphia, PA, Taylor and Francis, 1998.

20 Cozzani V, Tugnozi I, Salzano E. Prevention of Domino Effect: From Active and Passive Strategies to Inherently Safer Design. *Journal of Hazardous Materials.* A139, 209–219, 2007.

21 Gupta J, Edwards D. Inherently Safer Designs – Present and Future. *Trans Icheme.* 80, Part B, 115–125, 2002.

22 Hess G, Johnson J. Deconstructing Inherently Safer Technology. *Chemical and Engineering News*. 92, 1, 11–16, 2014.

23 Kemsley J. EPA to Require Chemical Companies to Consider Inherently Safer Technologies. *Chemical & Engineering News*. March 2, 2016. http://cenblog.org/the-safety-zone/2016/03/epa-to-require-consideration-of-inherently-safer-technologies/. Accessed March 27, 2018.

24 American Chemical Society. Inherently Safer Technology for Chemical and Related Industrial Process Operations. 2017. www.acs.org/content/acs/en/policy/publicpolicies/sustainability/ist.html. Accessed March 27, 2018.

25 Holt M, Andrews A. Nuclear Power Plant Security and Vulnerabilities. Congressional Research Service. 2014. https://fas.org/sgp/crs/homesec/rl34331.pdf. Accessed April 14, 2017.

26 U.S. Department of Homeland Security. Chemical Facility Anti-Terrorism Standards (CFATS). www.dhs.gov/chemical-facility-anti-terrorism-standards. Accessed March 27, 2018.

27 North American Electric Reliability Corporation. CIP Standards. www.nerc.com/pa/stand/pages/cipstandards.aspx. Accessed April 14, 2017.

28 Parsons, Inc. Cybersecurity Threat to Chemical/manufacturing www.parsons.com/media%20library/cybersecurity-chemical-industries.pdf. Accessed April 3, 2017.

29 Stoye E. Security Experts Warn Chemical Plants Are Vulnerable to Cyber Attacks. Chemistry World. www.chemistryworld.com/news/security-experts-warn-chemical-plants-are-vulnerable-to-cyber-attacks-/8632.article. Accessed March 27, 2018.

30 Sentryo. Is the Chemical Industry Targeted by Cyberattacks? www.sentryo.net/is-the-chemical-industry-targeted-by-cyberattacks. Accessed March 30, 2017.

31 Scholl H, Chatfield A. The Role of Resilient Information Infrastructures: The Case of Radio Fukushima During and After The 2011 Ester Japan Catastrophe. International Journal of Public Administration in the Digital Age. 2014. www.igi-global.com/article/the-role-of-resilient-information-infrastructures/112001. Accessed March 27, 2018.

32 Hollnagel E, Treiten C, Albrechtsen E. Resilience Engineering and Integrated Operations in the Petroleum Industry. 2011. www.researchgate.net/publication/282325239_Resilience_Engineering_and_Integrated_Operations_in_the_Petroleum_Industry. Accessed March 27, 2018.

33 Gortler L. Merck in America: The First 70 Years from Fine Chemicals to Pharmaceutical Giant. *Bulletin for the History of Chemistry*. 25, 1, 1–9, 2000.

34 Hawthorne F. *The Merck Druggernaut: The Inside Story of a Pharmaceutical Giant*. Hoboken, NJ, John Wiley & Sons, 2003.

35 Budavari S, ed. *The Merck Index: An Encyclopedia of Chemicals, Drugs, and Biologicals*. 11th ed. (Centennial Edition), Rahway, NJ, Merck and Co., Inc. 1989.

36 Merck & Co., Inc. Merck Announces Fourth-Quarter and Full-Year 2016 Financial Results. http://investors.merck.com/news/press-release-details/2017/Merck-Announces-Fourth-Quarter-and-Full-Year-2016-Financial-Results/default.aspx. Accessed May 2, 2017.

37 New Jersey Future. Smart Growth Leadership Award: Merck & Co., Inc. 2003. www.njfuture.org/smart-growth-101/smart-growth-awards/2003-award/. Accessed March 27, 2018.

38 Peterson M. A Company Move That Hasn't Irked Neighbors. *New York Times*. November 15, 1992. www.nytimes.com/1992/11/15/nyregion/a-company-move-that-hasn-t-irked-the-neighbors.html. Accessed March 27, 2018.

39 U.S. EPA. Region 2. Merck & Company, Integrated Report. www3.epa/gov/region02/waste/fs.merck.htm. Last updated 2/23/16. Accessed April 10, 2017.

40 NJDEP. Merck, Motiva/Shell Each Pay State Over $2 Million For Ground Water Claim. www.nj.gov/dep/newsrel/2006/06_0002.htm. Accessed April 21, 2017.

41 Greenberg M, Schneider D, Parry J. Brownfields, A Regional Incinerator and Resident Perception of Neighborhood Quality. *Risk: Health, Safety & Environment.* Summer, 241–259, 1995.

42 Greenberg M. *Restoring American's Neighborhoods. How Local People Make a Difference.* New Brunswick, NJ, Rutgers University Press, 1999.

43 Greenberg M, Schneider D. *Environmentally Devastated Neighborhoods.* New Brunswick, NJ, Rutgers University Press, 1996.

44 Berry M, Rondinelli D. Proactive Corporate Environmental Management: A New Industrial Revolution. *Academy of Management Perspectives.* 12, 2, 38–50, 1998.

45 Schering-Plough. 2009 Report on Safety, Health And Environment. Kenilworth, NJ, Schering-Plough. 2009.

46 Merck & Co, Inc. 2015/2016 Corporate Responsibility Report. Kenilworth, NJ. www.msdresponsibility.com/environmental-sustainability/. Accessed April 10, 2017.

47 Merck & Company. Merck Named to 2013 Working Mother 100 Best Companies. September 17, 2013. www.mrknewsroom.com/news-release/diversity/merck-named-2013-working-mother-100-best-companies. Accessed March 27, 2018.

48 Merck & Co. 5 Success Stories In Energy Management: Merck & Co. Inc. www.buildings.com/article-details/articleid/3574/title/5-success-stories-in-energy-management-br-merck-co-inc-. Accessed March 27, 2018.

49 Merck & Company. Merck Receives 2016 Energy Star Sustained Excellence Award From the U.S. Environmental Protection Agency, April 1, 2016. www.mrknewsroom.com/news-release/corporate-news/merck-receives-2016-energy-star-sustained-excellence-award-us-environmen. Accessed March 27, 2018.

50 Merck & Company. Merck And Codexis Honored With Presidential Green Chemistry Challenge Award For Novel Process For Sitagliptin Synthesis. www.fiercepharma.com/pharma/merck-and-codexis-honored-presidential-green-chemistry-challenge-award-for-novel-process-for. Accessed March 27, 2018.

51 Bassam E, Ruck R, Dienemann E, Emerson K, Humphrey G, Raheem I, Tschaen D, Vickery T, Wood H, Yasuda N. Merck's Reaction Review Policy: An Exercise In Process Safety. Safety Of Chemical Possesses. *Organic Process Research & Development.* 17, 1611–1616, 2013.

52 Baldwin C. Merck to Cut Jobs at NJ Facilities, Company Says. July 1, 2016. www.statnews.com/pharmalot/2016/07/12/merck-jobs-layoffs/. Accessed March 27, 2018.

53 Lowe D. Merck Cuts Back – Again. http://blogs.sciencemag.org/pipeline/archives/2016/07/12/merck-cuts-back-again. Accessed March 27, 2018.

54 Palmer E. Updated: Merck Manufacturing Shedding Another 2,600 Jobs. Pharma News. August 18, 2015 www.fiercepharma.com/supply-chain/updated-merck-manufacturing-shedding-another-2-600-jobs. Accessed March 27, 2018.

55 Carroll J. Merck Triggers a New Round of Layoffs In R&D Reorganization, Pushing More Jobs Into Cambridge, San Francisco. https://endpts.com/merck-triggers-a-new-round-of-layoffs-in-rd-reorganization/. Accessed March 27, 2018.

56 Mukherjee S. Merck to Axe Jobs as Part of New Research Focus. http://fortune.com/2016/07/13/merck-cuts-restructure-reserch/. Accessed March 27, 2018.

57 Merck & Co. Merck Opens New Manufacturing Facility In Hangzhou, China. April 16, 2013. www.mrknewsroom.com/press-release/corporate-news/merck-opens-new-manufacturing-facility-hangzhou-china. Accessed March 27, 2018.

58 Bojo J. Vietnam Development Report 2011: Natural Resources Management. Washington, DC, World Bank. 2011. http://documents.worldbank.org/curated/en/ 50919 1468320109685/Vietnam-development-report-2011-natural-resources-management. Accessed March 27, 2018.

59 The World Bank. World Bank Supports Managing Pollution in Vietnam's Most Industrialized Provinces. *The World Bank.* 2012. www.worldbank.org/en/news/press-release/2012/10/25/world-bank-supports-managing-pollution-in-vietnams-most-industrialized-provinces. Accessed March 27, 2018.

60 The World Bank. Vietnam Industrial Pollution Management Project. http://projects. worldbank.org/P113151/vietnam-industrial-pollution-management-project?lang=en&tab=overview. Accessed March 27, 2018.

61 The World Bank. Vietnam Industrial Pollution Management Project (P113151) Implementation Status & Results Report. *The World Bank*. 2017. http://documents. worldbank.org/curated/en/951081489775247076/pdf/ISR-Disclosable-P113151–03–17–2017–1489778846820.pdf. Accessed March 27, 2018.

62 VietNamNet Bridge. Vietnam Risks Becoming "Pollution Haven". *VietNamNet Bridge*. 2016. http://english.vietnamnet.vn/fms/special-reports/160868/vietnam-risks-becoming-pollution-haven-.html. Accessed March 27, 2018.

63 Covestro. CO_2 as a Raw Material. 2017. www.co2-dreams.covestro.com/en. Accessed July 7, 2017.

64 German Federal Ministry of Education and Research (BMBF). From Waste to Raw Material: Can CO_2 Replace Crude Oil in the Future? 2016. www.bmbf.de/pub/From_Waste_to_Raw_Material.pdf. Accessed July 7, 2017.

65 The European Chemical Industry Council (Cefic). Congratulations to the European Responsible Care Awards Winners 2016. 2016. www.cefic.org/Responsible-Care/Awards/Winning-Projects-From-Previous-Years/Awards-2016/#winners. Accessed July 7, 2017.

66 China Daily. Firm Considers Licensing Method to Transform Gas Into Plastic Products. *China Daily*. 2016. www.chinadaily.com.cn/business/2016–10/31/content_27224544.htm. Accessed July 7, 2017.

67 Thorpe L. Unilever's Factories Send Zero Non-hazardous Waste to Landfill. *Guardian*. 2015. www.theguardian.com/sustainable-business/2015/apr/30/unilevers-factories-send-zero-non-hazardous-waste-to-landfill. Accessed July 7, 2017.

68 Sustainable Brands. Unilever Achieves New Milestone, Collaborates to Fuel Global Zero-Waste Movement. *Sustainable Life Media*. 2016. www.sustainablebrands.com/news_and_views/waste_not/sustainable_brands/unilever_announces_latest_global_zero_waste_achievement. Accessed July 7, 2017.

69 Unilever. Unilever Announces New Global Zero Waste to Landfill Achievement. 2016. www.unilever.com/news/press-releases/2016/Unilever-announces-new-global-zero-waste-to-landfill-achievement.html. Accessed July 7, 2017.

70 Unilever. Rethinking Waste – Towards a Circular Economy. www.unilever.com/sustainable-living/reducing-environmental-impact/waste-and-packaging/rethinking-waste-towards-a-circular-economy/. Accessed July 7, 2017.

6 Concentrating locations at major plants (CLAMP)

Introduction

Chapter 6 has four objectives:

- Define the CLAMP policy and summarize its advantages and disadvantages;
- Review evidence about application of the CLAMP policy;
- Consider public reaction to the policy; and
- Illustrate the use of the CLAMP policy with current efforts to site a storage facility for elemental mercury in the United States.

Concentrating locations at major plants (CLAMP) as a policy

An organization is implementing a CLAMP policy when it chooses to build new noxious facilities at a site that it owns or otherwise controls and the site already has similar facilities. Traditional location theory focused almost entirely on adding new sites, now commonly called "greenfield" locations. However, as Chapter 3 showed, an explicit or implicit CLAMP policy was in effect at many noxious facility sites, such as the Manville asbestos site, the American Cyanamid chemical facility, Kin-Buc Landfill, and at other sites when managers expanded facilities in response to increasing markets. This chapter is slightly longer than the other case study chapters. That deliberate choice was made because while clamping has a long history, there is little formal literature on some of the issues that arise. Hence, I have added a few more cases and discussion to this chapter.

The advantages and disadvantages of a generic clamping policy for noxious facilities are summarized in Table 6.1.

An imperative for a developer is to try to avoid crossing a line that changes local public and government from a perception that an industrial or waste management facility is an asset or at least tolerable to a judgment that it has become a liability – in other words, destabilizing a functioning local social, political and economic system by adding more facilities that are not well tolerated.

The community can also destabilize an area by not managing the nearby land. Here are several examples. In 1986, the author and colleagues were asked

Table 6.1 Advantages and disadvantages of a CLAMP policy

Criteria	Advantage	Disadvantage
Raw materials and energy	Likely to be able to more easily and efficiently store and recycle on large CLAMP site	Large sunken investment that may be difficult to divest in the event raw materials and energy prices change
Labor	Should be able to schedule on-site skilled labor, train and update workers, and more readily set flexible schedules for employees	Local worker dependency will increase with large size and likely leave the organization with a long-term responsibility
Markets	Advantage to the organization when a CLAMP site is near to market	As markets shift the site may become inefficient to serve new markets
Capital	Less capital required for land acquisition, and likely less new infrastructure required	May need to demolish and remediate part of the site before proceeding
Land	Less land needed if site has space	On-site construction may impact surrounding areas and be considered unfavorably by local officials, publics, and other developers
Local cooperation	Advantage if there is a history of cooperation with local, state and regional governments and not-for-profits	Major disadvantage if there is hostility, especially from local and state governments
Infrastructure, including transportation	Major savings if there is excess capacity or existing capacity can be readily expanded	Problem if the existing infrastructure is not easily expandable and new units would increase on-site congestion
Agglomeration	Advantage if there are opportunities for concentrating activities and workforce	Expansion could exacerbate congestion and increase air and water quality issues
Environmental science and technology	An opportunity to remediate existing public and occupational-related problems, and to provide more sustainable and resilient conditions on one site rather than have to achieve these conditions at multiple ones	Failure to focus on local conditions at a CLAMP site are likely to lead to much more serious challenges at a later time
Public concerns and environmental justice	Some local population may want the expansion, and other sites do not have to deal with potential negative issues	Some local residents may oppose the policy on the grounds that the site already has facilities and should not bear the brunt of negative consequences, chronic exposures, and major hazard events, as well as social and political disruption
Local government autonomy	Influential local government officials may favor the expansion and press state, as well as federal and private organizations to support the project	Local officials may oppose the plan but may change their minds and new elected officials may be opposed

by the U.S. Nuclear Regulatory Commission (NRC) to assess why residential populations appeared to the NRC staff to be disproportionately increasing in towns hosting nuclear power plants. The NRC did not want local populations to increase because more people would need to be evacuated should there be a problem at the reactor.

Hence, the staff wanted us to independently evaluate their assessment about population change around nuclear power plants and explain why supposedly frightening places appeared to be attracting new residents. We[1] found that while the vast majority of the U.S. population did not want to live near a nuclear power plant, others were happy to do so because they were not worried about an accident. Also, they appreciated paying lower property taxes, and benefited from more public services than surrounding towns. The most important driving factor appeared to be that the utilities paid all their property taxes to a single local jurisdiction in some of these places, which attracted many would-be homeowners to these local areas to obtain tax and service benefits. In this situation developers were adding housing into places to take advantage of a population that was insensitive to common risk perceptions. The local government did nothing to dissuade the developers.

A notably different example comes from southern New Jersey. The author and a colleague[2] investigated the area surrounding one of the worst hazardous waste landfills in the United States. When we drove to the site, we saw young boys riding their bicycles up and down the slope of the landfill, which was covered by grass, and so it did not appear to them to be a major hazardous waste site. With further investigation we learned from their parents that they had been able to purchase homes adjacent to the site for 20–30% less than the established market rate when people left the area to escape the site. The declaration of the site as a major hazardous waste facility had frightened people and caused many to sell their properties as soon as possible at discounted prices. The local government did not stop them. Neither of the situations described in the two previous paragraphs was a good outcome for protecting human health and safety.

The objective is to try to avoid crossing lines that destabilize areas. Len Fisher's[3] book addresses the issue of how we can use science to identify what he calls "Crashes, Crises, and Calamities" – what I called crossing a line where an asset becomes a problem. Fisher identifies indicators that suggest something bad is about to happen, that is, some new critical path with positive feedback is being created that includes the following:

- Concentration of stress at weak points;
- Potential for runaway effects;
- Loss of resilience;
- Loss of existing patterns; and
- Inability of cope with the added stress.

Fisher's thesis makes sense for engineered systems. However, Fisher's logic is relevant to thinking about a CLAMP policy as a likely destabilizing mechanism.

An organization proposing a CLAMP policy should be in contact with local citizens groups, study media stories, talk with elected officials and to other sources to determine if what they are thinking about is already causing or is likely to cause undue local economic, social, and political stress, as well as pose a threat to human health and the environment.

During the late 1990s, I suggested scanning to the U.S. Army when it was proposing to build incinerators at eight continental U.S. locations where chemical weapons were stored and were to be destroyed. This meant building one or more noxious facility incinerators next to stored chemical weapons. The army retained retired reporters at four of the eight locations to try to determine community concerns and how these could be addressed. It also conducted public surveys, held public meetings and took other actions. The reporters helped identify issues, some of which did not come out in the public meetings and surveys. The goal was to avoid destabilizing the community because in the end the stored weapons would be destroyed, which would reduce local risk and increase property values. But during the destruction process, we were concerned that the public would focus on fears associated with the incineration process rather than the goal of destroying these local and potentially deadly stockpile of weapons.

In my experience, the chance for hostile pressure to build up is heightened when a community believes that it is repeatedly targeted for facilities that no one else wants. In Chapter 4, I briefly noted my experience with the Chester, Pennsylvania environmental justice case, which reached the U.S. Supreme Court. The bottom line in that case was that the plaintiffs argued that the area already had far more than their share of environmental hazards, and they framed their argument in the context of cumulative risk. Their law suit presented evidence that an exceedingly large share of noxious facilities were located in one city and in one census tract in the city. The suit also presented data about high age-adjusted death rates, infant mortality rates and presented other evidence of high risk in the local population. The claim was supported by the U.S. EPA but not by the State of Pennsylvania.[4] In the end, their case was convincing and the plaintiffs were granted the right to sue under the Title VI of Civil Rights Act of 1964 by the Federal court. But the applicant withdrew the request to build the facility, which made the court ruling to allow the suit to move forward irrelevant.[4]

While the Chester case was not ruled on by the U.S. Supreme Court, the case illustrates that the public was provoked into a law suit by what it felt was an endless series of new noxious facility locations, which they strongly believed increased cumulative human health and safety risk in the area. The people were frustrated with the private company proposing the site, but much of their anger focused on some of its local officials and the state government for ignoring NAIMBY claims. The community members I interacted felt demeaned and were not going to sit by while their area became a sacrifice zone.

Not every clamping proposal takes the form of the Chester one, or others that I am familiar with. An arguably ethically challenged clamping option is to shoehorn a new facility into a site like Chester without reducing its emissions, nor increasing safety and security, jobs, taxes and other local benefits, nor to offer to negotiate local benefits. This implies more human health, environmental and other local costs without benefits to the neighbors. A more fathomable approach for many would be to knock down an older unwanted and noxious plant and replace it with a new up-to-date facility, even if it is bigger, and negotiate with the community about a variety of issues. Variations exist between these poles.

Evidence of a CLAMP policy

Early nuclear examples

The idea of a CLAMP policy as will be shown below has existed for more than a century. One of the first statements was by Burwell, Ohanian and Weinberg.[5] In 1979, focusing on the nuclear power industry and writing in *Science*, they called for an "existing-site" policy. Forecasting growth to 1000 large (~1000 megawatt-electric) power reactors, these Oak Ridge engineers examined the implications of locating 1000 units on 100 sites versus 500 sites. The enumerated many advantages of the 10 per site option, such as:

- Minimize the search for new land;
- Limit contaminant spread from multiple sites;
- Reduce duplication of security forces;
- Limit the need to move spent nuclear fuel;
- Limit the need to do environmental studies of potential new sites;
- Ease the burden of reactor decommissioning;
- Building and retaining a skilled workforce and organizational capacity; and
- Reduce public opposition and increase support.

In essence, their existing site policy called for a set of permanent sites for nuclear power equivalent to those for managing DOE's defense nuclear operations at Hanford, Savannah River, Oak Ridge, and others. The authors asserted that some concentrating was already occurring and that many of existing large nuclear power sites could accommodate many more facilities. The paper presents no disadvantages of clamping.

Studies exploring the advantages and limitations of clamping nuclear power sites, while focusing on the benefits, also include disadvantages. For example, I reviewed a study of siting four nuclear power plants in Ocean County, New Jersey.[6] Ocean County is located in central New Jersey, about 85 miles southwest of New York City, 55 miles east of Philadelphia and in the heart of what became an affluent suburban area (Figure 6.1) A variety of economic benefits would have occurred for the utility. However, there were issues, including the reality that Ocean County was the fastest growing county in New Jersey between 1960 and 1970 and one of the fastest growing in the United States.

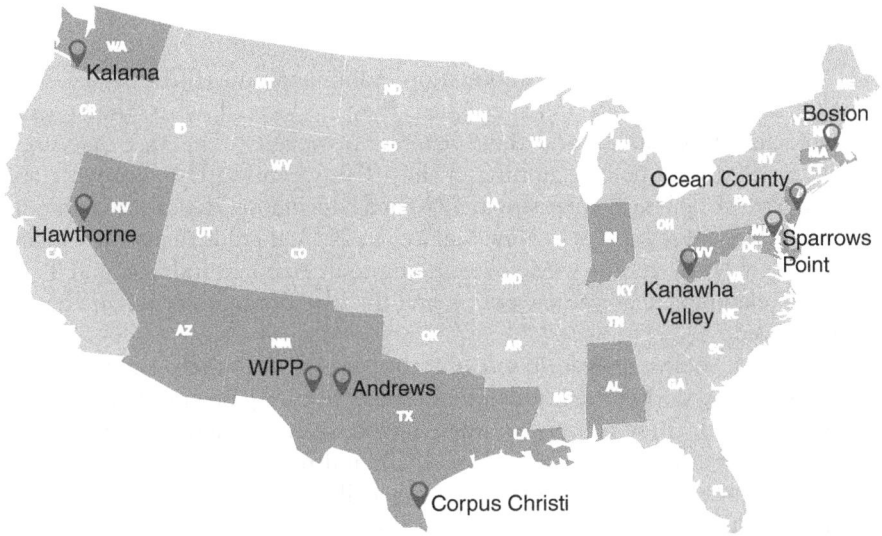

Figure 6.1 CLAMP case study locations.

The population grew by more than 90%! A clustered set nuclear power plants was never built.

Other studies, more theoretical in nature than the Ocean County one, explored clustering 40 nuclear power plants, each with 1200-megawatts electric, including fuel recycling centers.[7] Other studies proposed smaller clusters and proposed building nuclear power plants in the ocean (see Chapter 8).

Clamping of coal, natural gas, nuclear, and oil facilities producing electricity

In 2015, we measured the evidence for clamping nuclear power, oil, coal, and natural gas plants in the United States. Specifically, we compared the number of plants and sites in 1990 and 2013 in five states: Arizona, Indiana, New Jersey, New Mexico, and Texas.[8] These were selected because the authors were familiar with the sites and from experience knew if sites were clamped or not, rather than relying on names and addresses, which can be misleading.

The five states illustrate three regional patterns[8] (Figure 6.1):

- Arizona, New Mexico, and Texas are adjacent southwestern states with a wide variety of types of energy sites. The states have experienced massive population and economic growth;
- Indiana is a Midwestern state with a historical focus on fossil fuel and modest economic and population growth;

- New Jersey is a Northeastern state with slow economic and population growth, and heavy reliance on nuclear, gas and renewables.

We used the U.S. Energy Information Administration (EIA) database that includes number of units and sites where they are located, when they opened, when some were closed, and their generation capacity. The vast majority of units in 1990 were built beginning in the 1950s when the U.S. economy massively expanded. The average site in 1990 had had almost 3.6 units, with a range of 4.2 in New Jersey to 2.9 in New Mexico. In 2013, there is slightly less clamping in New Jersey and New Mexico, and more in Arizona, Indiana, and Texas. Overall, clamping is not a new policy, and the idea of ten units per site has not materialized.

However, these aggregate results do not present an entirely clear picture of what happened during the last quarter-century. Focusing on sites build between 1991 and 2013, there is a major difference between gas plants and the others. Across the five states, 74% of new coal, oil, and nuclear plants were clamped. Only 20% of gas plants were clamped, typically next to or replacing old coal plants. Multiple gas plants are being inserted at new natural gas facility sites, and now there is clamping at those new sites.

Texas, experiencing massive economic and population growth, especially in the triangular-shaped area with Dallas–Fort Worth in the north, Houston in the southeast and San Antonio in the southwest has added many units and has been clamping many of them. New Jersey exemplifies a fossil fuel poor area that heavily relies on nuclear power. Until the Clean Air Act took force in the early 1970s, New Jersey used coal to produce about two-thirds of its electricity. Now, about half comes from four nuclear power plants and the other half from natural gas. While the author has not visited every site, many of the old coal sites are being used for natural gas plants.

Overall, electricity-generating facilities historically were market-oriented, located near urban-industrial centers. With the development of a transmission grid, they are less confined. Clamping prevails, but less than some believe to be possible and desirable.

Clamping of petrochemicals

The market for petrochemicals derived from oil, gas, and coal is much more likely to be international in a globalizing world than it was a generation ago. Petrochemical facilities produce benzene, butadiene, ethylene, methanol, propylene, toluene, and xylene, which can then be converted to high value consumer products. For example, while I was writing this chapter, I played with my two young grand-daughters. In short order, we had balloons, multiple plastic toys, and plastic dolls on the floor; we sat on plastic patio furniture, and ended by putting most of the toys into a large plastic bag for storage.

Clamping petrochemicals in the United States is fascinating because U.S. industry has markedly changed in less than a decade from a net importer to an exporter of gas and oil. Readers who are old enough will remember that in 1973 the United States' petroleum supply was cut off. Energy independence became a goal of both political parties. In early 2009, a quarter century after the oil embargo, an official U.S. government report indicated that the United States would continue to be a major importer[9] and that many new facilities, especially liquefied natural gas (LNG) would be needed to bring natural gas to the United States.

This longstanding expectation abruptly changed when fracking of shale began to produce large supplies of oil and gas. Asbury[10] summarized the impact of this change on the location of the petrochemical industry. Writing in January 2017, she notes that the United States crude oil production had more than doubled during the last five years. Domestic fuel prices had fallen, and U.S. manufacturing industry had taken advantage of these prices to regain some of the manufacturing losses described in Chapter 1. In less than a decade the U.S. has become the largest world exporter of oil and gas, and is expected to become the third largest exporter of liquefied natural gas (Qatar and Australia export more).

With regard to petrochemical plants, Asbury asserts that the petrochemical industry has been adding small units at existing sites rather than building large new sites. Large new sites, she asserts, are considered high economic risk because of gas and oil price volatility, the complexity of developer and contractor interactions required for planning and executing new projects, and lack of experience in project planning and implementation because of retirements and job migration.

Asbury's report lists petrochemical projects underway and approved and others in the planning stages. For example, two tables list 25 U.S. projects intended to produce plastics. Twenty-four of the 25 are at existing sites – in other words, CLAMPs. The geography is clustered but not surprisingly with 13 in Texas, seven in Louisiana, and one in Alabama, reflecting the growth of supply from oil and gas producing facilities in the area and shipped to the area. Three are in Pennsylvania, the eastern state noted for shale oil and gas (Figure 6.1).

The lone exception among the 25 was a new greenfield project by M&G Polyethylene in Corpus Christi, which would be the largest plant of its kind in the world.[11] Photos and pictures show that it does not to fit my definition of a CLAMP policy site. Purchased in 2012, the site is adjacent to navigable water, to three railroads, and has available utilities and access to large raw material supplies. It would fit the old "greenfield" designation drawn to resources and transportation access.

With regard to methanol sites, the report lists 13 projects underway or approved, of which 10 are in Texas or Louisiana. Two are in West Virginia and one was proposed for the State of Washington. From photos and reports, five are CLAMP sites and eight are new sites. Each of the eight new sites has an interesting story to tell about economic risk and uncertainty. For example, the Lake Charles, Louisiana, methanol site is part of a U.S. federal government effort to demonstrate that carbon dioxide that otherwise would be emitted into the atmosphere can be recovered and turned into a raw material.[12] The idea is to recover the CO_2, ship it west to Texas and pump it into the ground to enhance oil recovery in fields entering the final stages of their economic use. The U.S. government is to provide up to a $2 billion loan for this $3.8 billion project.

A new plant targeted for St. James Parish in Louisiana also has interesting circumstances. Barnes[13] indicates that the $1.85 billion plant is designed to be an entry point for Yuhuang Chemical's entry into this potentially lucrative business in the U.S. The Chinese company recognizes that it may take time for it to become profitable. Indeed, it hopes to expand the site to three facilities. Both of these greenfield Louisiana projects are long-term investments, one to demonstrate a technology and the second to establish a foothold in a market.

Some of the other new site proposals include political interventions aimed at reducing government-related uncertainty. Two methanol plants in West Virginia involved cooperation among local and state officials and California-based Methanol California.[14] The idea is to use Marcellus shale gas to cluster one or two methanol plants in the industrialized Kanawha Valley.[14] Notably, part of the siting process is that company plans to take apart two methanol plants in Brazil and Slovenia and reassemble them at these West Virginia sites. The one that I was able to review in some detail is located near an existing Dow Chemical site and along a gas line and is characterized by the developer as a small and opportunity to quickly move into a strong market. Here again, the classic draw of resources and transportation are paramount, and reading between the lines local and state cooperation, as well as business acumen seem apparent.

Not all the new sites have been favorably greeted by public groups. Northwest Innovation proposed a new site at the Port of Kalama to convert natural gas to methanol and ship it to China to be used to produce plastics. While Washington Governor Jay Insell is said to have contributed to the idea while on a trip to China, environmental groups and some local officials have opposed the proposal for this greenfield site.[15] The plant was granted state permits in June 2017, and the permit was challenged by environmental advocate groups.

Overall, there are relatively few new site proposals for petrochemicals, and these few with a few exceptions are located in areas near gas or oil pipelines, and have superior transportation and infrastructure assets. They represent the kind of sites that traditional location analysis would have picked a half century ago, as well as today. Furthermore, with a few exceptions they are primarily at remote locations that will not gain negative attention from local communities. Like the CLAMP sites, which constitute the vast majority of petrochemical facility additions or planned additions, they tend to be located near sites that are already dedicated to petrochemical facilities.

Without doubt, LNG sites are the most interesting and controversial petrochemical facilities to site. During the first decade of the twenty-first century, with the United States still assumed to be a natural gas supporter, companies proposed LNG sites along the Eastern Seaboard near Boston (already had one), New York City, Philadelphia, in southern New Jersey near Delaware, and

Baltimore.[9,16] The economic logic was obvious. The potential markets were massive and pipelines connecting them already existed. However, in the face of public, local and state official opposition, almost every one of these Atlantic Ocean sites was scraped, even those that received go ahead from the Federal Energy Regulatory Commission (FERC).

One proposed in the Baltimore area is illustrative. For many years, the Bethlehem Steel company had a major integrated steel mill at Sparrows Point Maryland, located about 15 miles east of the City of Baltimore (30 minutes by automobile) (Figure 6.1). Sparrows Point was the largest steel producing plant in the world. When Bethlehem Steel closed, the site lay idle until the AES Corporation proposed a LNG facility at the site. The location was opposed by the fishing industry, by the Governor of Maryland, and other elected officials, including local officials. I read every comment that was part of the environmental impact statement, and support was not to be found. One of the strongest comments was made by a local fire chief who feared that his company would not be able to put out an LNG fire.[16] AES continued to pursue the idea, and did receive all but one permit. But AES stopped the process, and did not explain why. My interpretation is that during the period that it pursued the site, the U.S. became a natural gas exporter, and the site had been designed for importing not exporting.

FERC[9] listed 21 LNG sites in late 2009. Four were approved and under construction and 17 were approved but not under construction. Fourteen were Gulf Coast locations, and nearly all of these are CLAMP sites near an existing LNG export terminal, an oil refinery, or another chemical plant. Nearly all the others were along the Atlantic Ocean and one was to be in Oregon.

By early 2017, three CLAMP sites in the Gulf Coast were constructed, as was a LNG terminal in Massachusetts east of Gloucester, which is located northeast of Boston and juts out into the Atlantic Ocean (Figure 6.1) The Massachusetts sites were not operating at the time I wrote this chapter. The other sites Atlantic coast and one in Oregon were opposed by state government officials. Without a strong domestic market need, the CLAMP sites in the Gulf Coast are much more likely to be built than those off the Atlantic Coast, which face much stiffer political opposition. Furthermore some of the LNG sites in the Gulf are expected to be located 25 to 60 miles from the coast, which may lessen potential local political and public concern.[17]

Arguably, the current failure of the new Atlantic Ocean off-shore sites is attributable to public opposition, much as there has been opposition to west coast sites.[18] However, companies may reactivate some of these applications and efforts if and when the shale gas industry becomes less lucrative, in which case the debates about LNG safety and other impacts will surely re-emerge. On the other hand, if the coal industry, which some have written off as dead, were to revive in the U.S., then there would be less need for LNG. It is not difficult to see that analysts are now living in a world that is not only perilous because of public and some government officials' perceptions of noxious industry but also because the demand for specific resources is so difficult to estimate with any reasonable certainty. It follows that small adjustments, especially clamping make a great deal of sense to reduce uncertainty.

Public reaction to a CLAMP policy: nuclear facility cases

Public surveys about energy sources almost always target a representative sample of people from a nation or state. Few directly or even indirectly are a test of a CLAMP policy. The result is that while we have many studies asking national audiences about preferences for nuclear, coal, gas and other sources, we lack data sets representing the views of people who live in host communities.

Rosa's[19,20] excellent surveys of the United States population are typical. For example, in regard to nuclear power plants, in 2001 and 2004, Rosa observed that the majority of U.S. residents did not want a nuclear power plant in their jurisdiction. Of course, very few had one in their jurisdiction, so their views were in response to a hypothetical facility or one that they may have seen while driving nearby or from an airplane. Their major reasons for opposition were fears about the safety of the nuclear reactions, distrust of the nuclear industry and its government managers and concerns about nuclear waste disposal.

An MIT survey directly tied together new nuclear power plants with waste management, with about two-thirds of U.S. respondents to a year 2007 survey indicating that they would be more supportive if waste storage issues were reduced.[21] In the United Kingdom, Poortinga et al.[22] found that 42% of respondents were opposed to new nuclear power plants and 34% supported the idea. Greenberg's review of many surveys shows that public support reflects their national government policy position.[23] That is, if the national government supports nuclear energy (e.g., France), the public supports it. If the national government does not, then the public does not (e.g., Austria). This observation has a built in feedback loop with government officials and the public supporting and strengthening each other's positions. National surveys are important for companies and government analysts, but do not help us understand public reaction to the idea of clamping new nuclear sites.

The most complete data set on a clamping policy for any source of fuel for electrical energy has been collected by Bisconti for the Nuclear Energy Institute.[24-26] Using random digit dialing phone surveys of 18+ year old adults, they have collected data focused in areas with ten miles of commercial nuclear power plant sites in the United States. For example, their 2007 survey[24] was of 1152 adults who lived within a 10-mile radius of 64 nuclear power plant sites. The authors reported that 79% felt well or somewhat well informed about their local nuclear power plant; 76% of respondents agreed to adding a new nuclear reactor at the site; 86% had a "favorable" reaction to their existing nearby plant; and 71% agreed with storing used nuclear fuel at the local site until it can be safely moved to a permanent disposal facility.

In regard to the CLAMP policy, the most important data compare those living near sites with the national population, the vast majority of whom do not live within 10 miles of a site. Bisconti's 2015 survey of plant "neighbors" consisted of 1080 responses (60 sites × 18 people per site = 1080).[26] In order to avoid the issue that results are overly influenced by respondents that work at the site or have a family member that does, this group of respondents were excluded from the sample.

Three findings are particularly noteworthy:

- 50% of plant neighbors strongly favored and 33% somewhat favored the use of nuclear energy as one of the ways to provide electricity in the United States. This compared with 27% and 42%, respectively, in the United States as a whole.
- 69% of neighbors would accept a new nuclear power plant in their areas. This support was 79% in the West, 73% in the South, 69% in the Midwest, and 58% in the Northeast;
- Disproportionately, plant neighbors agreed with reasons commonly given for favoring nuclear power, such as reliability, efficiency, job creation, clean air, and energy security.

The Bisconti et al. series began during the 1980s. The authors conclude that support for nuclear power has increased since 1996, and with regard to a CLAMP policy the authors assert the data show a reverse of NIMBY near nuclear power sites, in other words, nearby residents demonstrate stronger support for the plants than those who do not live nearby. In fact, their observation agree with those in Chapters 2 and 3 that local support was the anticipated norm a half century ago, and with earlier comments in this chapter that some people do not fear noxious facilities and value any local benefits that they provide, and choose to move into communities with the facilities.

Given the fact that the Nuclear Energy Institute has supported the Bisconti et al. studies, the U.S. Department of Energy (DOE) was hesitant to assume that these findings hold for its nuclear defense sites. They supported this author's efforts to study public perceptions, values, trust, preferences, economic impacts, and support for a CLAMP policy at DOE sites.[27–34] A survey conducted in 2008 of 2701 people is the most relevant to clamping. Among the respondents, 2101 of them resided within 50 miles of 11 existing major defense nuclear waste sites and/or a nuclear power plant; and 600 resided elsewhere in the United States. We asked them to choose between approval of a CLAMP site, a new facility elsewhere, and no facility.

The national sample is the base. Among these 600 national respondents, 34% favored CLAMP for new nuclear power plants, 52% for new nuclear waste management facilities, and 50% for new nuclear laboratories. Table 6.2 shows that their responses are almost identical to those who live within 50 miles of one of the nuclear sites. The CLAMP option for each type of site increases among residents of the host county and those who work at the site or have a family member that does.

In addition to living in a host county, advocates for clamping disproportionately were male, white, affluent, college-educated, and among those who had

Table 6.2 Public preference for CLAMP policy for nuclear facilities, 2008, %

Population sampled	Nuclear power favor			Nuclear waste management favor			Nuclear laboratory favor		
	CLAMP**	Other	None	CLAMP	Other	None	CLAMP	Other	None
National sample	34	31	35	52	18	30	50	22	28
Aggregate of nuclear power and defense waste sites	34	31	35	53	19	27	50	22	28
Host counties	40	29	31	55	21	24	58	19	23
Worked at site or family member worked at site	50	29	21	62	20	18	61	22	16

Source: adapted from Greenberg.[27]

Notes
* The 11 nuclear-site-centered areas were near Calvert Cliffs (MD), Palo Verde (AZ), multiple plants in south Texas, North Anna (VA), Nine Mile Point (NY), and the waste management sites near Hanford (WA), Idaho Falls (ID), Oak Ridge (TN), Savannah River (SC), and Waste Isolation Pilot Plant (NM).
** "CLAMP" means that they favored a new site clamped at their local site; "Other" means a new site but not near them; and "None" means that they did not want a new facility.

visited the site. These respondents trusted nuclear managers; they were not worried about nuclear technologies or other local environmental issues. None of these findings were surprising, as they mirror previous research findings about the nuclear industry and the role of trust and familiarity.[35–41] Also, they were worried about continuing to use coal because they link it to climate change.

Another interesting CLAMP-related finding is that the four DOE former defense sites had the highest degree of support for clamping more nuclear waste management facilities: Hanford (61%), Idaho National Laboratory (67%), Oak Ridge (60%) and Savannah River (52%) (see Figure 4.1). This study did not allow me to learn exactly why this was the case. Yet, speculating, these four have existed for more than a half-century, and they have advisory boards that frequently meet with the DOE about a range of issues including land use, which is a key issue in the clamping process. Their local populations have strongly supported national defense, and when this survey was conducted we estimated that 14% to 19% of their local economies depended on the DOE site, and their regions viewed the sites as critical to their immediate future, especially if the sites replaced an economic void.[30,32–33,42]

If siting a nuclear power plant and nuclear waste management facilities is challenging, Chapter 7 will show siting a repository for nuclear waste is fraught with multiple levels of challenges. Yet, the public sometimes is receptive to siting proposals. In 2009, the DOE proposed building what they called energy parks at some of their former defense sites.[43] Their idea was to figure out the science for managing climate change by leveraging the science and engineering communities at these sites. At a talk, I supported this plan, arguing that this idea was a chance to turn bitter lemons for many people into lemonade. The parks were expected to include facilities that would work on the following:

- Renewable energy, focused on solar, wind, biomass, and geothermal energy;
- Fossil fuels, especially cleaner coal, improving gas turbines;
- Electricity generation, transmission, distribution;
- Hydrogen generation;
- Emission controls, and carbon sequestration; and
- Nuclear power science and engineering, fuel cycle and waste management.

Some would consider these facilities to be noxious, others would not. By definition they are noxious because many of the materials and products are hazardous. In the end, the DOE chose not to implement this policy. However, the idea was sufficiently intriguing that I conducted a survey of residents living within 50 miles of six DOE defense sites and a national sample for comparison. The national sample found 29% supported the facility in their state. This compared to over 40% in the DOE regions near the Hanford, Washington and Waste Pilot Isolation Pilot Plant in New Mexico, and most notably 54% in the host counties for the six DOE defense sites.

Case study: storing and managing elemental mercury

Mercury (Hg) is a basic metal in the periodic table. Mercury has the atomic number 80, and the unusual arrangement of its electrons has made it an important element. It is the only metal that is liquid at room temperature. My high school chemistry teacher Mr. Romano demonstrated this property to our utter amazement by placing two drops of mercury on a bench and watching it form a droplet. Then he added a third and was able to merge them together into a single larger one. This mobility property led to Mercury's wide use in thermometers, barometers and blood pressure monitors. Mercury is a good conductor of electricity, and accordingly has been widely used in electrical relays and florescent lights. Mercury has various less obvious but noteworthy applications, for example in explosives and Mercury was widely used in batteries.

Mercury forms amalgams with many other metals, notably gold and silver. Consequently, it was widely used in dental amalgams and in recovery of gold and silver from mines. Mercury was used in medical and cosmetic applications. For example, Mercurochrome was antiseptic and during my childhood applied to small cuts and scratches. The list of applications of mercury uses is long, and goes back to ancient Egypt, China, and Rome. In short, Mercury in its various forms has been in use for over 3500 years.

Mercury, however, needs to be carefully managed because many of its compounds are extremely toxic.[44,45] Epidemiological studies demonstrate tremors, sleep disturbance, and impaired cognition in workers who have had chronic exposure to mercury vapors. Acute exposures of as little as eight hours have produced impaired pulmonary functions. Acute exposures to mercury vapors can cause serious central nervous system effects, including hallucinations, delirium, and even suicidal tendencies, as well as excitability and irritability.

The World Health Organization (WHO) the U.S. Occupational Safety Health Administration (OSHA) and National Institute of Occupational Safety and Health (NIOSH) have set occupational exposure limits for Mercury, and the United States Environmental Protection Agency regulates environmental emissions. Occupational health effects have been one of the major reasons why governments have tried to control Mercury. In addition, mercury spills into the environment have produced some of the most tragic results. Fish and shellfish concentrate Mercury in their bodies, typically as methyl mercury, which is highly toxic. Tuna, tilefish, swordfish and other predators eat smaller species and in the process concentrate Mercury. This biomagnification process can produce Mercury levels of 10 times or more above lower level fish.

Eugene and Aileen Smith's book *Minamata*[46] published in 1975 tells a story of the poisoning of fishermen who developed lips and limb tingling, numbness, slurred speech and lost motor functions. Some died. Investigations showed that an industrial company used mercury as a catalyst in manufacturing plastics, perfumes, drugs, and photographic chemicals. The process waste was dumped in the water, consumed by fish, biomagnified and then the fish were eaten by residents. This book's photographs of injured people are as painful to view as those of Nagasaki and Hiroshima after the nuclear detonations occurred. Another notable water pollution example was at what is now the DOE's Oak Ridge plant that produced lithium isotopes during the 1950s and 1960s, and as part of the United States elemental Mercury stockpile[47] (see below for more information).

While Minamata disease is the single worst cluster of cases we know about, studies systematically demonstrate that Mercury needs to be controlled. Organized by the United Nations Environmental Program (UNEP), 140 nations agreed to collaborate to prevent mercury emissions. In the U.S. the power plants that burn coal and gas account for more than 60% of emissions, and over 10% comes from gold production. The remaining sources are from non-ferrous metal smelters, cement production, and lye production.

In the United States, the EPA is the responsible agency for regulating much of the Mercury use and production. The largest users of Mercury have been finding substitutes for Mercury. For example, today's thermometers use digital electronics (usually for warmer) and dyed alcohol gases (usually for colder). U.S. states began to ban mercury thermometers in 2001 and a decade later it is difficult to obtain one for home use.

Before focusing on the role of United States Department of Energy and the elemental Mercury siting requirement, we note that EPA is the major responsible federal agency in the United States. Mercury wastes are classified as toxic under the Resource Conservation and Recovery Act, a hazardous substance in the Comprehensive Environmental Resource, Compensation, and Liability Act, and a toxic pollutant under the Clean Water Act and Clean Air Act.

The Mercury Export Ban Act of 2008

(a) The DOE's role in locating a facility resulted from the Mercury Export Ban Act of 2008. After October 14, 2008, no United States federal agency is permitted to sell or distribute any elemental Mercury (with a few specified exceptions). The act prohibits the export of Mercury effective January 1, 2013. The ban on exports and then on exchanging elemental Mercury among federal agencies means that an inventory of about 10,000 or more metric tons need to be stored.

(b) The U.S. Department of Energy (DOE) is required to find a site to store elemental mercury. Table 6.3, taken from DOE's Mercury EISs, shows the sources of elemental Mercury that the DOE is expected to store.

The quantity of elemental Mercury to be stored from the gold mining and recycling facilities is uncertain. The gold mining amount depends on the price of gold in the United States. If the price increases, more Mercury will be used in gold mining. Recycled Mercury comes from the phase-out of Mercury use in thermometers, switches, thermostats and many other applications. A good deal of that waste reclamation has already been completed, so uncertainty about this form of Mercury is lower than for others.

Table 6.3 Anticipated Mercury inventory for the DOE to store*

Source	Years sent to storage facility	Metric tonnes
By-product from coal mining	2013–2052	3700–4900
Waste reclamation and recycling	2013–2052	2500
DOE Y-12 National Security Complex, Oak Ridge Tennessee	2013–2014	1200
Closure of chlor-alkali plans or conversion to non-Mercury-cell technology	2013–2019	1100
Total		8500–9700

Notes
* Data in table from United States Department of Energy (pp. 1–4).[48]
Not included in this table is approximately 4400 metric tonnes that the United States Army is preparing to store at the Hawthorne Army Depot in western Nevada.

In regard to Mercury from the chlor-alkali process, there are three methods to produce chlorine and caustic soda (sodium hydroxide). One of these takes advantage of Mercury's alloying ability. This process, however, is being phased out and the DOE is being asked to store the recovered Mercury from that process.

Elemental Mercury at the DOE's Y-12 facility in Oak Ridge is the fourth major source. Between 1950 and 1963, about 11,000,000 kg of Mercury was used at the Oak Ridge facility for lithium isotope separation and for several other purposes. About 3% was estimated to be admitted into the air, soil, rock and East Fork Popular Creek was heavily contaminated.[47–51] DOE will recover whatever it can from the Oak Ridge facility for storage.

Siting

Arguably, all of this elemental Mercury can still be used. However, in essence it is being treated as if it were a toxic waste product, that is, it is to be confined in a remote location without any obvious future use. In this regard, it is akin to "spent" nuclear fuel from commercial nuclear reactors, which, in fact, can be recovered and used to manufacture new fuel, but currently is being treated as a waste product (see Yucca Mountain below in Chapter 7).

DOE states the objectives of the long-term mercury management and storage program as follows:

- Protect human health and the environment and ensure safety of the public and facility workers.
- Meet the requirements of the Mercury Export Ban Act of 2008.
- Comply with applicable Federal, state, and local statutes and regulations.[48]

(pp. 1–5)

The U.S. environmental impact statement process has been critical as a process tool for evaluating how to manage elemental mercury, as it has been for noxious substances as a whole.[16,48–50] It is the most widely used federal government process to assess the siting of noxious facilities, and must be used when the site is government owned or requires government permits and other permissions.[16,48–50]

Public Law 91–190 – the National Environmental Policy Act (NEPA), which became the law on January 1, 1970 – contains a requirement for an EIS. The broad objectives of the law were to create and maintain conditions man and nature exist in "productive harmony" both in the present and future. Title one, section 102 (2)c required all federal agencies to prepare an EIS on any "… proposals for legislation and other major federal actions significantly affecting the quality of the human environment." (42U.S.C. 4332). Section 102c specified that the five following elements be covered:

1   The environmental impact of the proposed action;
2   Any adverse environmental effects which cannot be avoided should the proposal be implemented;
3   Alternatives of the proposed action;
4   The relationship between local short-term uses of the human environment and the maintenance and enhancement of long-term productivity; and
5   Any irreversible and irretrievable commitments of resources which would be involved in the proposed action should be implemented.

NEPA has been criticized by business and some government officials as taking too long to implement and costing too much money to complete. The other side of the argument is that it does not require directors of federal departments to follow EIS recommendations.[16] In short, an EIS requires analysis but the recommendations need not be followed.

In regard to facility siting, such as finding a site for a long-term mercury storage site, an EIS should choose locations and evaluate them in regard to the following possible impacts:

• The natural environment, such as, ecologically sensitive areas, endangered species, water quality, and wetlands;
• Human health, including both acute and chronic disease impacts, and during both construction and operation of the facilities;
• The built environment, for example, implications for local land use and zoning, economic development, historic properties, aesthetics and community organization;
• Environmental justice, specifically potential impacts of identified minorities and low income communities; and
• Transportation, for instance, noise and vibration, traffic congestion, pollutant emissions, parking, mass transit, pedestrian traffic.

The public is supposed to participate in a "scoping" phase when anyone is permitted to offer remarks in person, through the mail or via the web; and every inquiry must be answered. Further, in regard to siting, every location that has been considered must be evaluated in regard to the same criteria. Hence, at a minimum, an agency can evaluate one site and a no action alternative. The no action alternative is tracing the impacts of continuing existing practices. In most cases, the EIS team will start with multiple locations, eliminate some as infeasible, and focus their attention around two to four that seem possible. After the EIS is completed and evaluated, new ideas may be suggested and these are accommodated in supplements.

Few significant major government and private building projects in the United States do not engage the EIS process. An agency may argue that a large EIS is not required because the impacts are likely to be small. It can, instead, write an environmental assessment, which is a mini version of an EIS. An agency can also assert that a proposed project has so little potential for impact that no assessment is required. During the last 40 years, the number of requests and allowances of environmental assessments rather than full-blown EISs has substantially increased, as has the call for no assessment. Also, some federal agencies have argued that they should be excused from the NEPA process.[16] These are controversial developments, however, they are beyond the scope of this chapter because it is hard to think of a circumstance whereby a major waste management facility receiving any federal government resources for construction or transportation would escape the EIS process, although it is conceivable that the Congress, a U.S. President and the Courts would think otherwise in response to an emergency or threat.

Using the EIS process as its framework, in March 2009, the DOE requested expressions of interest in siting and hosting the Mercury storage facility. Private sector and federal government agencies were invited to respond, and the DOE requested its own sites to determine if they had suitable locations. The DOE did not preclude a greenfield option, but it is difficult to see how one would emerge from the list of seven siting criteria it listed:

1   The facility(ies) will not create significant conflict with any existing DOE site mission and will not interfere with future mission compatibility.
2   The candidate host location has an existing facility(ies) suitable for mercury storage with the capability and flexibility for operational expansion, if necessary.
3   The facility(ies) is, or potentially will be, capable of complying with RCRA permitting requirements (see Chapter 5, Sections 5.2.4 and 5.3[This refers to the EIS not this book]), including siting requirements.
4   The facility(ies) has supporting infrastructure and a capability or potential capability for flooring that would support mercury loadings.
5   Storage of mercury at the facility(ies) is compatible with local and regional land use plans, and new construction would be feasible, as may be required.
6   The facility(ies) is accessible to major transportation routes.
7   The candidate location has sufficient information on hand to adequately characterize the site.[48]

(pp. 1–6)

All but number 3 are relatively straightforward. In regard to point number 3, the Resource Conservation and Recovery Act of 1965 (as amended) sets forth criteria for the storage of hazardous waste that limit the number of places that can host a site, including for location, design and construction, and a permit is required to operate such a site. These permits take time to acquire, and consequently, the DOE would logically look at sites that already had permits for storage of elemental mercury and/or other hazardous materials. In order to meet the federal requirements the site would need to be designed with proper spill containment features and emergency response procedures; there would need to be security and access controls; fire suppression technologies; ventilated storage and handling areas; the building would need to be fully enclose and protected from the weather and would need a reinforced concrete floor able to accommodate Mercury, a very heavy material.

In response to its request, the DOE received responses from 10 companies and federal agencies (Table 6.4).

The reader should not take these locations at face value. The actual sites are not necessarily at these addresses. One of the three sites (Veolia) withdrew its application. Meritex's site proposal was for an underground storage facility in a former limestone mine. The DOE concluded that this location would be difficult to permit. Lowland did not propose a specific site, as required by the request.

As required by law, the DOE held public meetings at the seven possible locations over a three-week period beginning July 21 and ending August 13, 2009. The EIS reports that 300 people attended these meetings and 507 documents were submitted. Some of these documents were submitted via a toll-free fax, U.S. mail and the web.

Table 6.4 Basic data from organizations responding to the DOE's request*

Site	Location	Status
Grand Junction, Disposal Site, DOE (public)	Grand Junction, Colorado	Reviewed
Hanford nuclear site, DOE (public)	Richland Washington	Reviewed
Hawthorne Army Depot (public)	Hawthorne, Nevada	Reviewed
Idaho National Laboratory, DOE (public)	Idaho Falls, Idaho	Reviewed
Bannister Federal Complex's Kansas City plant, DOE (public)	Kansas City, Missouri	Reviewed
Lowland Environmental Services (private)	Knoxville Tennessee	Rejected
Meritex Enterprises (private)	Cumberland Furnace, Tennessee	Rejected
Savannah River nuclear site, DOE (public)	Aiken, South Carolina	Reviewed
Veolia ES Technical Solutions (private)	Henderson, Colorado	Rejected
Waste Control Specialists (private)	Andrews Texas	Reviewed

Note
* Data summarized from United States Department of Energy.[48]

The comments can be organized into distinct areas:

- Site selection;
- Alternative storage opportunities;
- Transportation (estimating 50 to 100 truck shipments a year to the site);
- Health and safety;
- Facility accidents;
- Possible land-use impacts;
- Environmental impacts;
- Environmental justice;
- Sociological impacts;
- Cultural resource and Native American issues;
- Regulatory compliance concerns.

Space does not permit an accounting of all the comments. Instead, I focus on those that directly bear upon site selection. Respondents gave the following reasons to oppose one or more of the selected sites:

- In competition with existing site missions, including cleanup activity;
- Private ownership;
- Proximity to surface and/or groundwater;
- Natural hazards, such as earthquakes.

Some respondents supported several of the candidate sites for the following reasons:

- Proximity to existing mercury storage sites;
- Large site with available space;
- Remote location.

Not surprisingly, the media took notice. For example, NBC news ran a story that listed the seven sites and underscored the opposition from officials from Colorado, Idaho, South Carolina, Nevada, and Kansas.[52] A site near Andrews, Texas, was not opposed, nor supported by local officials. A subheading in the NBC news story was "best option in Texas?" A Texas state official was paraphrased as indicating that the private company already had permits to receive a certain amount of mercury at the site. In January 2010, the governor of Idaho[53] supported the site near Andrews, Texas. Governor Otter of Idaho stated "[he] will not allow Idaho to become the nation's dumping ground for its elemental Mercury. If they want to put it in a desolate and useless place, they should put it on the (U.S.) capital grounds." Comments from the other elected officials were somewhat less colorful and certainly more respectful, but clearly indicated no interest in hosting the site in their state.

In 2011, the DOE expressed a preference for the site in Andrews, Texas, area where Waste Control Specialists already store other hazardous wastes, including low level nuclear. The area is right on the Texas–New Mexico border[49] (Figure 6.2). The justification for the Andrews site on the one hand is easy to understand.

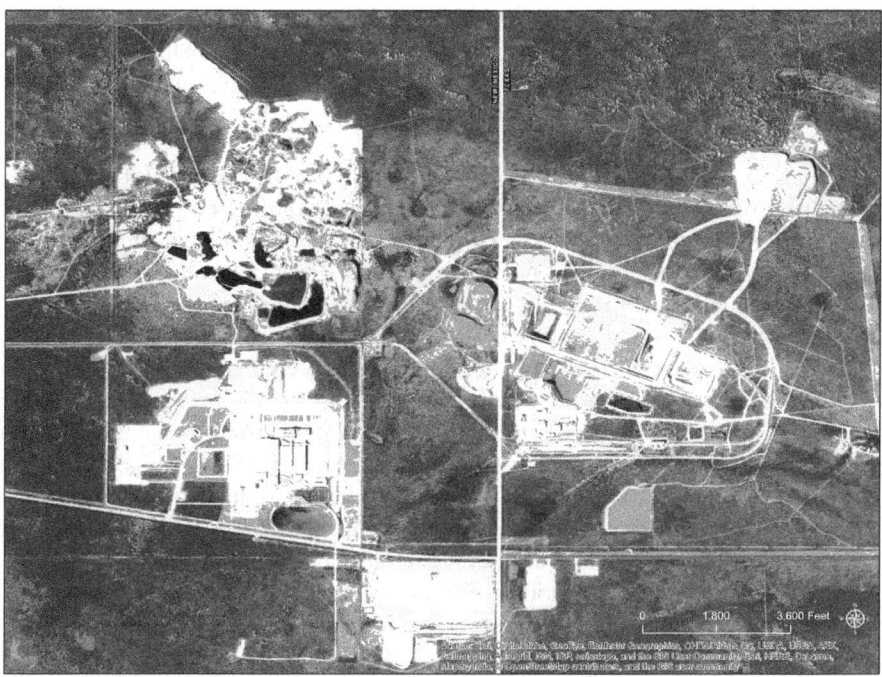

Figure 6.2 Texas–New Mexico border waste sites.

The site itself is remote, and the analysis of all the data led the DOE to conclude that potential impacts across all of the criteria discussed above would be either "negligible" or "minor." A flavor of why this site was chosen as illustrated by the following quotations from the EIS:

> habitat in the immediate vicinity of the construction sites consists mainly of shrub lands with grassy patches.... Since an existing building would be used for interim mercury storage, no land would be disturbed and no terrestrial resources would be impacted.[49]
>
> (pp. 4–125, 126)

> no wetlands or aquatic resources exist within the proposed location of the new Mercury storage facility at WCS. Therefore, no impacts on wetlands are aquatic habitats are expected.[49]
>
> (pp. 4–126)

> there would be no impact on historic resources from Mercury storage facility construction or from operations. Surveys conducted found in absence of historic occupational exploitation of the area.[49]
>
> (pp. 4–126)

The EIS continues with the words no, negligible, and minor impacts. The reality is that this is a very isolated site with almost no receptors of value to humans. In 2010, Andrews County had a population of less than 15,000 people, and a population density of 9/square mile (3.34/km^2). It takes about 40 minutes to drive to Odessa, Texas, and about 50 minutes to drive to Midland, Texas, which is about 45 miles away.

Not only is it remote but it is also geologically stable, dry and until relatively recently one of the few locations in the United States there would be public support for the storage of hazardous nuclear waste. I found that at the time of the EIS the nearest residence to the proposed site was 3.4 miles (5.4 km) away. The risk to members of the public is classified as "negligible."

On the other hand, a number of the other candidate sites had similar profiles. The obvious exception is the facility located about eight miles south of Kansas City. Another one of the locations that had public support was at the WIPP locations. WIPP has an underground storage facility for transuranic nuclear wastes, and is a major siting success story, including increasing support from the public for the activity.[42] Two of the three sites are immediately outside the current WIPP boundary and the third is within it. However, in 2014 there were two accidents below ground at the WIPP site, and consequently it is not yet clear what the public's reaction to the idea of hazardous waste storage is at this time. Hence, it is probably a prudent decision to pick the Texas site for Mercury storage.

In September 2013, the DOE[50] submitted a final EIS with a supplement that included three additional sites near the DOE Waste Isolation Pilot Plant (WIPP) near Carlsbad, New Mexico. That EIS supplement reiterated the selection of the Andrews, Texas, site. Waste Control Specialists use a 1338-acre (541 hectares) to treat store and landfill a variety of hazardous chemical and radioactive wastes. It is located approximately 31 miles (50 km) west of Andrews, Texas, and six miles (10 km) east of Eunice, New Mexico. The mercury storage facility would be fit in the property and an existing building could already store several thousand tons. The site is licensed by the United States nuclear regulatory commission to dispose of low-level radioactive waste.

This Congressional-driven process is a fascinating exercise in inter-agency and public-private partnership. The DOE has the charge to find the site. But EPA is a major player in the Mercury toxicity, and by not precluding a privately owned site. The DOE faced the possibility of picking a site that is not strictly a CLAMP site. Yet, history tells us that should a privately owned site be chosen, the DOE, or another federal agency will need to step in when and if the company is no longer able to meet its mandate. For example, the failure of the New York State sited West Valley New York nuclear waste facility to economically succeed, ultimately led to shifting the responsibility to both New York State and the DOE

As of late April 2017, the DOE had not made a final decision about Andrews or another site. Several challenges remain at the Andrews site, despite local support. The most pressing is that normally the DOE announces the record of decision independently from the acquisition process. But the Andrews site is private, and hence the DOE would need to acquire a long term lease on land within the Waste Control Specialists property. Identifying the site in a ROD ahead of the acquisition process could be viewed as prejudicing what is supposed to be a full and open competition. Congress has provided the DOE with a budget to prepare the documents required for a ROD.

Another potential site makes some sense to this author, which is Hawthorne, Nevada. Hawthorne is a 226 square mile army depot with several thousand bunkers that hold ammunition and other DoD materials. Run by a contractor for the DoD, it has the advantage of a long history. The DoD Defense Logistics Agency stores National Stockpile Mercury in bunkers upgraded for this purpose. Yet, it has some limitations, most notably the fact that it is not identified as a hazardous waste storage faculty under the Resource Conservation and Recovery Act (RCRA). The DoD contractor would need to obtain an exemption in order to store elemental mercury from non-DoD sources at the site, which could be difficult.

Overall, as of late 2017, the DOE is making progress toward finding and implementing the plan to move elemental mercury to a remote site in a place already performing the same or a similar function. The Andrews Mercury site would be clamped to other hazardous metal and low level nuclear waste sites in the area.

International examples

There are other interesting international examples of building new noxious facilities at a site that already has similar facilities. Taiwan hosts the world's largest coal fired power plant, which is operated by the state-run Taiwan Power Company (Taipower).[54,55] Taichung power plant was initially constructed with four units in 1992, and four more units were added in 1996–1997. After the most recent expansion in 2005–2006, two more units were installed in addition to four gas turbines.[55] While meeting 20% of Taiwan's electricity demand, Taichung is the world's highest CO_2 emission releasing power plant,[56] which some have linked to increasing lung cancer rates in Taiwan.[57] Future expansions will consist of renovations on environmental facilities to cut nitrogen oxide and sulphur dioxide, in order to comply with Taiwan's tightening environmental regulations in the last decade.[58,59]

Belchatow Power Plant is Europe's largest lignite-fired power station and provides 20% of power demand in Poland.[60,61] It was initially commissioned with 12 units in 1982. This power plant is also known for its high CO_2 emission levels. It was named as the worst polluting power station by World Wildlife Fund in 2007 and by Sandbag Climate Campaign in 2009.[61,62] If you view the site from photos, it appears massive, with numerous units, stacks and cooling towers surrounded by flat land with a few farms and degraded land. The area looks devastated. In 2011, an expansion project was planned, which involved modernizing old blocks and constructing a new unit that will meet

European Union emission standards. In exchange for a loan for the expansion project, two outdated production units were to be closed down, but The European Commission ignored this agreement.[63] Later on, Belchatow was awarded funding to start CO_2 capture and storage project in order to reduce emission levels, but it was cancelled due to company's failure to secure necessary financing on their own share.[64]

Sinopec's, $8.8 billion expansion project of a petrochemical plant in Ningbo, China faced strong opposition in 2012.[65] The main reasons for protests were capacity expansion and production of Paraxylene (PX), a potentially harmful chemical used in manufacturing polyester, paints, and plastic bottles.[66,67] Despite efforts to reassure residents by implementing latest pollution-control technologies and planning to relocate 9800 houses away from the refinery site, protests started when farmers blocked a road near the plant. Involvement of thousands of students and middle-class residents grew the protests in the first couple days.[65] As a result, Ningbo government announced that they cancelled plans for paraxylene plant and decided to carry out scientific research for other parts of the project.[65,66] Protesters' distrust of local government was also reported since decision-makers did not follow through on promises and are said to have used such verbal declarations just to ease tension in the past.[65,67] A year later, the Ministry of Environmental Protection of the People's Republic of China provided local authorities approval power for expansion projects of paraxylene plants to simplify procedures for investors.[68] NAIMBY and NIMBY movements have been increasing in China, especially for industrial projects that threaten public health, local lifestyle, and culture.[66,69]

Final thoughts

Clamping is an old policy for noxious sites and should become even more prevalent. Business and government recognize the difficulty of finding a location, and the economics of adding on to an existing site allows them to avoid investing in a new site that might not be justified by uncertain markets in uncertain political environments. Clamp sites may not even make it onto the business pages and the public media, which these days is another benefit.

References

1 Greenberg M, Krueckeberg D, Kaltman M, Metz W, Wilhelm C. Local Planning v. National Policy: Urban Growth Near Nuclear Power Stations In the *United States, Town and Planning Review*, 57, 225–238, 1986.
2 Greenberg M, Schneider D. *Environmentally Devastated Neighborhoods: Perceptions, Realities, and Policies*, Rutgers University Press, New Brunswick NJ, 1996,
3 Fisher L. *Crashes, Crises, and Calamities: Using Science to Anticipate Disasters*. New York, Basic Books, 2011.
4 Greenberg M. *Explaining Risk Analysis: Protecting Health and the Environment*. New York, Earthscan/Routledge, 2016.
5 Burwell C, Ohanian M, Weinberg M. A Siting Policy for an Acceptable Nuclear Future. *Science*. 204, 1043–1051, 1979.

6 Meier P, Morell D. *Issues in Clustered Nuclear Siting: a Comparison of a Hypothetical Nuclear Energy Center in New Jersey with Dispersed Nuclear Siting.* Brookhaven, NY, National Center for Analysis of Energy Systems, 1976.

7 Nuclear Regulatory Commission, Office of Special Studies. Nuclear Energy Center Site Survey. 1975. https://catalog.hathitrust.org/Record/003164466. Accessed March 8, 2017.

8 Greenberg M, Coon M, Campo M, Whytlaw J. Chapter 15. Finding Locations for Endurably Objectionable Energy-Related Facilities: The CLAMP Policy, 234–246 in S Bouzarovski, M Pasqualetti, V Castan Broto, eds. *Energy Geographies.* New York, Routledge, 2017.

9 Federal Energy Regulatory Commission. Approved North American LNG Import Terminals, 2009. http://ferc.gov/industries/lng/indus-act/terminals/lng- approved.pdf. Accessed March 10, 2017.

10 Asbury M. Petrochemical Update. U.S. Downstream Engineering, Construction & Maintenance Market Outlook, 2017. http://(29e63718_efd4_4cfc_8955_2b12df9331 c8)_4780_19Jan17-US-Downstream-Engineering-pdf. Accessed March 8, 2017.

11 M&G PET Plant, Corpus Christi, Texas, United States of America. www.chemicals-technology.com/projects/mg-pet-plant-corpus-christi-texas/. Accessed March 27, 2018.

12 Osborne J. Feds Putting $2 Billion Behind Louisiana Methanol Plant with Carbon Capture. *Fuel Fix.* December 21, 2016. https://marcellusdrilling.com/2016/08/us-methanol-confirms-mdn-rumor-2-or-more-plants-coming-to-wv/. Accessed March 27, 2018.

13 Barnes S. St. James Methanol Plant on Track for Fourth Quarter Goundbreaking. June 8, 2016. www.businessreport.com/article/st-james-methanol-plant-track-fourth-quarter-groundbreaking. Accessed March 27, 2018.

14 Marcellus Drilling Com. U.S. Methanol Confirms MDN Rumor- 2 (or more) plants coming to WV. *Marcellus Drilling News.* http://marcellusdrilling.com/2016/08/us-methanol-confirms-mdn-rumor... Accessed March 10, 2017.

15 Bernton H. Kalama Methanol Plant Proposal Draws Challenge from Environmental Group. *Seattle Times.* www.seattletimes.com/seattle-news/environment/kalama-methanol-plant-proposal-draws-challenge-from-environmental-group/. Accessed March 27, 2018.

16 Greenberg M. *The Environmental Impact Statement After Two Generations: Managing Environmental Power*, New York, Routledge, November 2011.

17 McCulley R. for OE Digital. 2010. Deepwater LNG Double on the Horizon. www.oedigital.com/regions/item/23-deepwater-lng-double-on-the-horizon. Accessed March 27, 2018.

18 Boudet H, Ortolano L. A Tale of Two Sitings: Contentious Politics in Liquefied Natural Gas Facility Siting in California. *Journal of Planning Education and Research.* 30, 1, 5–21, 2016.

19 Rosa E. *The Future Acceptability of Nuclear Power in the United States.* Paris, Institute Français des Relations Internationales, 2004.

20 Rosa E. Public Acceptance of Nuclear Power: Déjà vu all over again? *Physics and Society.* 30, 2, 1–5, 2001.

21 Ansolabehere S. *Public Attitudes Toward America's Energy Options: Insights for Nuclear Energy.* MIT-NES-TR-08, Cambridge MA, MIT, 2007.

22 Poortinga W, Pidgeon N, Lorenzoni I, et al. Public Perceptions of Nuclear Power, Climate Change and Energy Options in Britain; Summary Findings of Survey Conducted During October and November 2005, Understanding risk working paper 06–02 citeseerx.ist.psu.edu/viewdoc/download?doi=10.1.1454.5644&rep=rep1. Accessed August 4, 2017.

23 Greenberg M. *Nuclear Waste Management, Nuclear Power and Energy Choices: Public Preferences, Perceptions, and Trust.* New York, Springer Publishers, 2012.

24 Bisconti Research, Inc. National Survey of Nuclear Power Plant Communities. for Nuclear Energy Institute, July–August, 2007. www.nei.org/newsandevents/news releases/nuclearpowerplantneighborsaccept.html. Accessed November 26, 2007.

25 Bisconti Research, Inc. U.S. Public Opinion about Nuclear Energy, Report for Nuclear Energy Institute, Washington, DC, May 5–9, 2005.
26 Bisconti Research Inc. and Quest Global Research. (2015) 6th Biennial Survey of U.S. Nuclear Power. www.nei.org/knowledge-center/public-opinion. Accessed February 24, 2017.
27 Greenberg, Michael. NIMBY, CLAMP and the Location of New Nuclear-Related Facilities: U.S. National and Eleven Site-Specific Surveys, *Risk Analysis*, 29, 1242–1254, 2009.
28 Greenberg M, Truelove H. Right Answers and Right-Wrong Answers: Sources of Information Influencing Knowledge of Nuclear-Related Information, *Socioeconomic Planning Sciences*. 44, 130–140, 2010.
29 Greenberg M, Popper F, Truelove H. Are LULUs Still Enduringly Objectionable? *Journal of Environmental Planning and Management*. 55, 713–731, 2012.
30 Frisch M, Solitare L, Greenberg M, Lowrie K. Regional Economic Benefits of Environmental Management at the U.S. Department of Energy's Major Nuclear Weapons sites, *Journal of Environmental Management*. 54, 23–37, 1998.
31 Greenberg M, Isserman A, Frisch M, Krueckeberg D, Lowrie K. Mayer H, Simon D, Sorenson D. Questioning Conventional Wisdom: The Regional Economic Impacts of Major U.S. Nuclear Weapons Sites, 1970–1994. *Socio-Economic Planning Sciences*. 33, 3, 183–204, 1999.
32 Greenberg M, Lewis D, Frisch M. Regional Economic Impacts of Environmental Management of Radiological Hazards: An Initial Analysis of a Complex Problem. *Environmental Planning and Management*. 44, 3, 377–390, 2001.
33 Greenberg M, Lewis D, Frisch M, Lowrie K, Mayer H. The U.S. Department of Energy's Regional Economic Legacy: Spatial Dimensions of a Half Century of Dependency. *Socio-Economic Planning Sciences*, 36, 109–125, 2002.
34 Greenberg M, Weiner M, Kosson D, Powers C. Trust in the U.S. Department of Energy: a Post-Fukushima Rebound. *Energy Research & Social Science*. 2, 145–147, 2014.
35 Pew Research Center. *Deconstructing Distrust: Americans View Government*. Washington, DC, Pew Research Center, 1998.
36 Poortinga W, Pidgeon N. Exploring the Dimensionality of Trust in Risk Regulation, *Risk Analysis*. 23, 5, 961–972, 2003.
37 Finucane M, Slovic P, Mertz CK, Satterfield T. Gender, Race, and Perceived Risk: the "White Male" Effect. *Health, Risk & Society*. 2, 159–172, 2000.
38 Flynn J, Slovic P, Mertz CK. Gender, Race, and Perception of Environmental Health Risks. *Risk Analysis*. 14, 6, 1101–1108, 1994.
39 Kahan D, Braman D, Gastil J, Slovic, P, Mertz CK. Culture and Identity-Protective Cognition: Explaining the White Male Effect in Risk Perception. *Journal of Empirical Legal Studies*. 4, 3, 465–505, 2007.
40 Weart S. Fears, Fantasies and Fallout, *New Scientist*. 136, 34–37, 1992.
41 Nemich C. Boise State survey shows statewide support for INL. news.boisestate.edu/newsrelease/032006/0302INLrelease.html. 2006. Accessed August 26, 2006.
42 Jenkins-Smith H, Silva C, Nowlin M, deLozier G. Reversing Nuclear Opposition: Evolving Public Acceptance of a Permanent Nuclear Waste Disposal Facility. *Risk Analysis*. 31, 629–644, 2011.
43 Greenberg M. Energy Parks for Former Nuclear Weapons Sites? Public Preferences at Six Regional Locations and the United States as a Whole, *Energy Policy*, 38, 5098–5107, 2010.
44 Zahir F, Rizwi S, Haq S, Khan R. Low Dose Mercury Toxicity and Human Health. *Environmental Toxicology & Pharmacology*. 20, 2, 351–360, 2005.
45 Renzoni A, Zoni F, Franchi E. Mercury Levels Along the Food Chain and Risk for Exposed Populations. *Environmental Research*. 77, 2, 68–72, 1998.
46 Smith WE, Smith A. *Minamata*. New York, Holt, Rinehart & Winston, 1975.

47 Brooks S, Southworth G. History of Mercury Use and Environmental Contamination at the Oak Ridge Y-12 Plant. *Environmental Pollution.* 159, 1, 219–228, 2011.

48 United States Department of Energy. Draft, Long-term Management and Storage of Elemental Mercury. DOE/EIS – 0423D, January 2010.

49 U.S. Department of Energy. Final Long-Term Management and Storage of Elemental Mercury. Supplemental Environmental Impact Statement. DOE/EIS-0423. September 2011.

50 United States Department of Energy. Final, Long-term Management and Storage of Elemental Mercury. DOE/EIS – 0423 – S1, September, 2013.

51 Burger J, Gochfeld M, Powers C, Kosson D, Clarke J, Brown K. Mercury at Oak Ridge: Outcomes from Risk Evaluations Can Differ Depending upon Objectives and Methodologies. *Journal of Risk Research.* 17 9, 1109–1124, 2014.

52 NBC News. Where to store 17,000 tons of mercury. U.S. lists seven possible sites, but local officials aren't interested. www.nbcnews.com/id/32125796/ns/us-news-environment/where-store-tons-mercury. Accessed July 22, 2014.

53 Kuraitis J. Idaho Governor Otter likes Texas – for Mercury storage. NewWest. January 29, 2010. https://newwest.net/city/article/idaho_gov_otter_likes_texas_for_mercury_storage/C108/L108/. Accessed March 27, 2018.

54 Pentland W. World's 39 Largest Electric Power Plants. *Forbes.* 2013. www.forbes.com/sites/williampentland/2013/08/26/worlds-39-largest-electric-power-plants/#3377340b58da. Accessed July 20, 2017.

55 Power-technology.com. "Giga" projects – the world's biggest thermal power plants. 2013. www.power-technology.com/features/feature-giga-projects-the-worlds-biggest-thermal-power-plants/. Accessed July 26, 2017.

56 Taipei Times. Taichung power plant world's worst polluter: survey. *Taipei Times.* 2008. www.taipeitimes.com/News/biz/archives/2008/09/04/2003422196. Accessed July 20, 2017.

57 The News Lens. Lung Cancer Cases Increasing in Taiwan and Medical Community Calls on People to Demonstrate against Air Pollution. *The News Lens.* 2015. https://international.thenewslens.com/article/32127. Accessed July 20, 2017.

58 Taiwan Power Company. Taiwan Power Company Sustainability Report 2014: A powerful future for you. 2014. www.taipower.com.tw/UpFile/CompanyENFile/2014.pdf. Accessed July 20, 2017.

59 Power Online. MHPS Receives Order from Taiwan Power Company for Coal-Fired Boiler and Environmental Facilities Renovation Project at Taichung Thermal Power Plants Units 1–4. *Power Online.* 2016. www.poweronline.com/doc/mhps-receives-order-from-taiwan-power-taichung-thermal-power-plants-units-0001. Accessed July 20, 2017.

60 ENS. Europe's 'Dirty 30' Power Plants Impair Climate Progress. *ENS.* http://ens-newswire.com/2014/07/22/europes-dirty-30-power-plants-impair-climate-progress/. Accessed July 25, 2017.

61 Macalister T. Meet Belcha – Europe's biggest carbon polluter (and it's about to get even bigger). *Guardian.* 2009. www.theguardian.com/environment/2009/jul/22/europes-biggest-carbon-polluter-coal. Accessed July 25, 2017.

62 World Wildlife Fund. *Dirty Thirty: Ranking of the most polluting power stations in Europe.* Belgium, Brussels. May 2007. wwf.panda.org/?100140/Europes-Dirty-30.

63 Climate Action Network Europe. EU extends life of Europe's biggest polluter while IPCC calls for coal phase out. *Climate Action Network Europe.* 2014. www.caneurope.org/publications/press-releases/723-europe-s-most-polluting-coal-power-plant-reverses-commitment-to-reduce-pollution. Accessed July 25, 2017.

64 MIT Energy Initiative. Belchatow Fact Sheet: Carbon Dioxide Capture and Storage Project. https://sequestration.mit.edu/tools/projects/belchatow.html. Accessed July 26, 2017.

65 Jacobs A. Protests Over Chemical Plant Force Chinese Officials to Back Down. *New York Times*. 2012. www.nytimes.com/2012/10/29/world/asia/protests-against-sinopec-plant-in-china-reach-third-day.html?_r=0. Accessed July 20, 2017.
66 Ruwitch J., Stanway D. China struggles for solution to growing NIMBY movement. *Reuters*. 2012. www.reuters.com/article/us-china-environment-idUSBRE8A01L020 121101. Accessed July 20, 2017.
67 Waldmeir P., Hook L., Anderlini J. Ningbo protest, response both typical of China's environmental debate. *Washington Post*. 2012. www.washingtonpost.com/world/asia_pacific/ningbo-protest-response-both-typical-of-chinas-environmental-debate/2012/10/29/ac4c8e5e-21f6-11e2-8448-81b1ce7d6978_story.html?utm_term=.9480d49cf303. Accessed March 27, 2018.
68 Jing L. Fears over expanded paraxylene projects as Beijing decentralises approval process. *South China Morning Post*. 2013. www.scmp.com/news/china/article/1378798/fears-over-expanded-paraxylene-projects-beijing-decentralises-approval. Accessed July 20, 2017.
69 Guardian. Chinese protest over chemical factory. *Guardian*. 2012. www.theguardian.com/world/2012/oct/28/chinese-residents-protest-chemical-factory. Accessed July 20, 2017.

7   Negotiating

Introduction

Chapter 7 has four objectives:

- Review major types of negotiations;
- Explain why negotiations should be pursued by parties involved in siting a noxious facility;
- Describe the facility siting credo guidelines and suggestions about how to use them as part of negotiations; and
- Compare the use of negotiations in the case of the Waste Isolation Pilot Plant and Yucca Mountain.

The advantages and disadvantages of a generic policy of negotiations for noxious facility are summarized in Table 7.1.

Negotiations

Individuals negotiate to reach an agreement about important issues. Some negotiations are about clarifying misunderstandings and take a few minutes to resolve. We have all been part of those negotiations. Others are about firmly held values and perceptions that prove to be unresolvable.

Organizations negotiate about benefits, prices, regulations, salaries, schedules, and many other issues, including siting facilities. Organizational negotiations can involve one-on-one meetings, and they can also involve scores of people on multiple sides with representatives meeting to negotiate specific parts of a large set of issues with or without mediators and arbitrators.[1] The form of negotiation can impact the results, especially when the parties expectations are not met. For example, I was a technical mediator for the American Arbitration Association (AAA)[2] when AAA was branching out into mediation and arbitration of environmental cases. During this period, I learned just enough to know how little I knew about the differences between forms of mediation and arbitration. Mediators are there to try to bring the parties together and do not make decisions. Arbitrators normally enter after mediation fails to resolve one or more

Table 7.1 Advantages and disadvantages of negotiations from the perspective of the proposing organization

Criteria	Advantage	Disadvantage
Raw materials and energy	Negotiations could lead to innovation in resource use	Could be required to use more and less desirable resources in response to negotiations
Labor	Could lead to new loyal labor force	New labor force could become dependent on the organization for employment
Markets	Negotiations could lead to new nearby markets that had not been considered	Organization could lose valuable markets if negotiation reduces scale and/or planned on-site activities
Capital	Negotiations lead to elimination of innovations that improve the cost-effectiveness of the project	Negotiations lead to higher costs and taxes for the organization. Community dependence may follow from a negotiated project that may be costly in the long run
Land	May need to alter site plan leading to technological or other innovations that would be beneficial in the long run	May need to change plan, requiring more land purchases elsewhere and unplanned additions on the site
Local cooperation	Opportunity to develop a reputation as a cooperative organization or enhance an existing one with local groups	Failure to succeed in negotiation may jeopardize the organization's reputation
Infrastructure, including transportation	Opportunity to reassess infrastructure needs for the facility and surrounding area	Could require costly unplanned add-ons
Agglomeration	Opportunities for adding agglomeration economies as part of negotiations with community, for example, by adding new jobs, income and tax-creating activities	Pressure to reduce scope of project may lead to elimination of important agglomeration economies
Environmental science and technology	Negotiations that lead to environmental and health challenges likely to beneficial to the organization in the long run	Short term increases in science, engineering, testing, implementation are likely to occur that may be challenging and infeasible for the organization
Public concerns and environmental justice	Need to be prepared with EJ and social justice principles and demonstration of such actions from within the organization	Without proper preparation organization will be vulnerable in public setting, damaging the organization's reputation
Local government autonomy	Chance to develop strong local allies who will support the organization and testify to its support	Possible that enemies will be created during negotiations that will carry into a new political administration

issues, and arbitrators make decisions. Subtle differences exist among each of these broad categories. In actual siting cases, some parties do not understand what process is being used; they expect one process and get something else, which can jeopardize negotiations.

A great deal of responsibility rests on the shoulders of the parties, mediators, arbitrators, and other professional staff. Communication skills, specifically listening and watching are essential in order understand each party's priorities. I have found that what is written down and provided to the parties before meetings about sites too often obfuscates what is really important to participants, that is, what they write down is typically about technical competence, what they really are concerned about are values.

Then, there is the challenge of resistance to negotiating. Negotiating for some means surrendering deeply held values, methods and processes. Some parties would rather argue, hoping to convince their opponents that their position is more ethical, truthful, rational, and of course is supported by data. Only when they fail to persuade the other party, often creating angst in the process, do some figure out that they need to be more flexible. For instance, I was involved in a case where the two parties were disputing a water resources issue. Each had done a report and when I examined the two reports, I immediately realized that the cause of their different results was not bad data or even inaccurate interpretation. It was that they were using different methods, both out of date, for analyzing the problem. When I informed the professional mediator, he had me explain this to them. Neither took it well at first, arguing that their approach was tried and true. It was for them, but other methods had come along that were more effective at getting to the point of the dispute.

Parties bring goals, values, and perceptions to the table. So do mediators and supposedly impartial experts. I faced this in my own work, trying to separate my technical assessment from my values.

Pace is another important element of negotiation. Personally, I much prefer a slow and deliberate pace that allows parties an opportunity to consider information. I worked with professional mediators with different styles. Some pressed the parties hard trying to quickly solve the dispute and others seemed to me to be involved in a counseling session before getting down to reaching an agreement.

Siting noxious facilities can be a wicked challenge for all parties, defying simple win–win solutions. Some stakeholders feel dismissed and angry, looking for retribution while others find negotiations exhilarating.[3–6] Much of their reaction has to do with what value they attach to their stake in negotiations. Some bargain and others dig their heels in, and some quit negotiations for the following reasons:

- They found an easier path to achieving their objectives, such as a behind the scenes political process, or vigorous protests that gain the attention of important people;

- They need a quick decision, they had to move on to another solution and/ or they did not have the resources to do what was necessary to end with their preferred results;
- The other parties were not represented by individuals who had the power to decide for their organizations;
- They concluded that the other parties were negotiating in bad faith;
- They found that the results would compromise their organization's practices and/or reputation;
- They were not prepared to deal with parties from cultures that looked at siting in ways that could not be overcome by negotiations; and
- They lacked contacts in the area where the site was proposed and recognized that this was a major disadvantage in negotiations.

In class, I always tell the students that they need to have at least three plans, just in case the first and second plans fail. When it comes to siting noxious facilities, I would start with the expectation of negotiations, but not expect negotiations to lead to acceptable outcomes for them, at least at first.

Negotiating to site a noxious facility

Considering what I learned about location theory as a student (Chapters 2 and 3), the reality that site-related negotiations are common today is a remarkably positive change for those of us who believe in democratic processes. As advocates of negotiations, our collective objective has been to reduce the number of contentious issues down to the point that parties would realize that their vested interest is to settle.

The most consistent contentious issue, one that often is unstated or buried in a sea of facts about the details of the site, is legitimacy, that is, who should make the decisions, as shown by the following observations based on my work in siting cases. After listing them, I describe them in greater detail:

- The basic principle is supposed to be that the federal government has primacy over states and local governments, and yet that is not what happens;
- Government agencies even at the same level (e.g., federal) have a difficult time of agreeing on what should be done and who should play what roles;
- Interpretation of laws and rules by judges and staff are more important than what the laws and rules say;
- Different states and local governments have markedly different land use and siting laws, rules, regulations and practices;
- For-profit and not-for-profit organizations are masters at finding and exploiting weaknesses, especially inconsistencies, in written law and practice; and
- Community advisory panels with broad representation have been effective at working with inconsistent and unclear information if they are involved as early as possible and given tools and sufficient resources and power to insert influence.

Legitimacy and geographical scale

I realize that the above six bullets read like the venting of a frustrated practitioner. Indeed, they are and they stem from seeing differences between what is written down, how it is interpreted and what actually happens. The national government legitimately is responsible for promote and regulate economic development, and local government is to provide basic local services such as police, fire, education, and others, such as land use control. The states fall between these two in our federalist system.

As the country has grown and markets have expanded, land use powers inevitably have overlapped and different parts of the United States developed their own ways of managing land use.[7-11] There are good reasons for the federal government and state government to have primacy in key land use decisions, even though local government normally is ceded that power:

- Local governments can too narrowly focus on impacts on a single jurisdiction;
- Siting decisions regarding a noxious facility have the potential to impact multiple jurisdictions;
- Large projects, notably noxious facilities, require permits and financial investments from federal, state as well as local governments;
- Federal and state governments have technical expertise that should be valuable to inform decisions that impact multiple jurisdictions; and
- Federal and state governments are needed to prevent private organizations from exploiting a lack of power at the local government level.

The federal government may legitimately claim primacy, yet when it does local governments and sometimes states are frustrated, feel powerless, file law suits and vigorously protest. Many hostile reactions from local people are grounded in environmental and social justice (see Chapter 4). For example, I vividly recall being verbally abused by an articulate local resident about a nuclear power plant location. Thumping on her chest and then pointing at me, she stated she was not going to sit by and watch this proposed facility give her children cancer. She described the process as watching a three-ring political circus between levels of government that led to the public's values and preferences being ignored. Her anger was palpable as she systematically tore the process apart as illegitimate and argued for more attention to local context. Whereas her arguments were aimed at nuclear power plants, I have heard the same arguments in other siting cases.

Below, I briefly return to the siting of chemical hazardous waste facilities to illustrate how difficult this process is for everyone, that is, angry residents, analysts who are legitimately trying to help, project managers, and elected officials. Chapter 4 described the chilling impact of the Resource Conservation and Recovery Act (RCRA) of 1976 and amendments in 1980 and 1984 on the ability to dispose of chemical waste in landfills. RCRA was the national government's response to headlines about hazardous waste disposal failures at Love Canal (NY), the Kin-Buc Landfill in New Jersey (Chapter 3), Times Beach, a suburb of St. Louis Missouri, and Valley of the Drums near Louisville, Kentucky. The outraged and frightened population demanded a solution. RCRA defined hazardous waste, established a process to track it from cradle to grave, and most important for this book, RCRA made standard landfill designs unacceptable. The Superfund Amendments and Reauthorization Act (SARA) of 1986 focused on the remediation of abandoned hazardous waste sites.

As part of these laws and regulations, the federal government requested that every state find locations where hazardous waste could be managed. The carrot was that the federal government provided resources to states to clean up the Superfund sites. In 1986, the request to find sites to manage waste for the next 20 years was an imposing challenge. I believe that the states were coerced and faced enormous pressure to respond. This effort was largely unsuccessful. For example, researchers[9-11] studied more than 80 private applications for hazardous waste management sites. They found that only 7% were ever approved. This failure included novel ideas that featured negotiations in Massachusetts and Minnesota.

Was this federally driven, state-implemented process a failure? Based on finding new sites, it failed. Yet, the failure led EPA to rescind the requirement that every state should find sites, and instead the United States figured out that they needed to relieve the pressure by advocating for pollution prevention (Chapter 5).

New Jersey mini case study: finding chemical hazardous waste sites

The complexity of federal–state–local processes is illustrated by what happened in New Jersey, a state at that time with one of the largest chemical industries in the United States, along with Texas and Louisiana. When New Jersey began planning its effort to respond to the EPAs mandate for a plan and sites, there was good reason to believe that a state-led process was necessary and would be successful. The state's population wanted a solution. Its population voted for a bond to help the state get started on planning for the program; the first state in the United States to do so.[12] But the issue was could a site be found in the state and would the public agree to it?

The state assumed primacy. The context for state primacy was as follows:

- There were 567 local municipal governments in this state (now 565), the only state with a population density exceeding 1000 per square mile, in other words, it is hard to find a site in an isolated area.
- A belief existed that in a state with a history of strong local land use control that no local government would be willing to be considered for a site;
- If there were volunteers, they might not be good sites in regard to geology, soils, terrain, population distribution, water supplies, and other considerations.
- New Jersey has had a successful history of special districts based on environmental needs: the New Jersey Meadowlands in the northeast of the state,

the Pinelands in the south, and the Highlands in the northwest, which was possible in this case.

In 1981, the state government passed a Waste Facility Siting Act, which included a siting commission with objectives to develop a state-wide siting plan.[13-18] The Commission estimated that the state needed an 80-acre above-ground waste disposal facility and two incinerators. Next, the Commission, after extensive consultation across the state, laid the groundwork for a constraint mapping exercise by identifying 36 siting criteria, which included details about geology, soils, terrain, and a host of other considerations. Using this approach they eliminated all but 11 locations in the entire state.

When the 11 sites were released, elected officials, residents and some environmental groups attacked the results and the process. Their attack was supported by careful and well-funded research that picked holes in the work by showing that the data were not accurate at the specific locations.

The Commission argued that this it was critical to the future economy of the state to find these three sites and that they state should have primacy.[17] They added that in addition to the constraint mapping, additional on-site research would eliminate the possibility of an unacceptable site being chosen. The process failed. In the end, the EPA backed off its threat to withhold Superfund dollars, the state government backed off its mandate, judges generally sided with local right to legally contest the findings, and not a single jurisdiction seriously considered stepping forward, or at least none that I know about.

New Jersey's failed effort was not the product of people with evil intent; in fact, the reality was that the group consisted of well-intentioned elected officials, staff and residents who really cared about the state. While they did not find sites that were acceptable, their work drove business and government to pollution prevention faster than if disposal sites had been built. By the time the process was over, for not-for-profits, members of the public, elected officials, and even the most dedicated analysts realized that the siting approach was problematic in one of the most affluent states in the United States where no community that the process identified wanted to be known as the chemical hazardous waste capital of the state. This was not simply local NIMBY; this was statewide NIMBY and opposition from multiple parties. Boholm and Lofstedt,[10] and Lake and Johns[9] drew similar conclusions from their research, that is, these cases are so complex that to label them as a NIMBY reaction is a misrepresentation of reality, one that ignores the social and political bases of resistance to noxious sites.

The process had one other benefit in New Jersey. It increased the credibility of the researchers in regard to the value of equity. Many members of the public were surprised to find that the process had identified one site that was on the property of an extremely wealthy family. When that site was disclosed people attending a meeting about the sites openly laughed, including this author. The audience praised the staff for having the courage to list this site.

Would this process have had a better chance of succeeding if it had started with waiting for someone(s) to volunteer? I do not think so in this state. In

1980, responding to a federal mandate to find a site for low-level nuclear waste, New Jersey created a low-level radioactive waste disposal committee. It moved toward a process of voluntarism, looking for one 50-acre site with a 200-acre buffer. There was publicity about the voluntary process. Several towns mentioned that they might be interested, but in the end in 1998 there were no sites, and in 1998 the search stopped and New Jersey's low-level waste continued to go to Barnwell, South Carolina. This same result occurred across the United States – a great deal of work produced little in the way of new sites. Andrews, Texas, was an exception (Chapter 6).

Much has as made about economic incentives in chemical and nuclear siting processes. In this regard, many elected officials and vocal community members are insulted by the idea that government or a developer would try to offer them money to surrender their values and the quality of life (see Yucca Mountain survey results later in the chapter). These cases demonstrate how stressful siting noxious facilities for all the parties. They start with different values, perceptions, assumptions, and data and bringing them to an agreed upon solution is a rare accomplishment.

The facility siting credo

The facility siting credo, is a list of suggested guidelines developed by Kunreuther et al.[19,20] to aid regional and local governments in negotiations. They compiled the list from surveys of stakeholders in siting controversies. I have reproduced their major points and added three bullets about each. Please note that this is my interpretation of their work, which was presented over a quarter century ago and remains valuable guidance to negotiators (Table 7.2).

Case studies: Waste Isolation Pilot Plant (WIPP) and Yucca Mountain

I am unaware of a definitive study about how much time it takes to site a noxious facility. Nevertheless, many decades of experience and reading tell me that private industry's bottom-line prevents a long search period. Businesses develop siting and non-siting options, make choices and move on. Government has different objectives. In many countries, for example the United States and China, government owns an enormous amount of land, and is able to find sites. But when the site is for a noxious facility, government encounters flak that can shoot down the best ideas. Finding a long-term underground repository for high-level nuclear waste easily wins the contest for length of time from idea to opening a facility; one or two generations is considered a success, if it ever opens.

I am familiar with efforts in Finland, France, Germany, Sweden, the United Kingdom and many other countries (see brief examples below) to find a repository for high-level nuclear waste from power plants, and have been briefed about efforts in China. However, I know a great deal more about efforts in the United

States, and the United States to date has had the biggest single success and failure. WIPP and Yucca Mountain are the poles. Hence, the focus of this chapter is on these two, especially WIPP, and the role of negotiations between the federal, state and local government, and the public in the siting processes.

High-level nuclear waste is a problem because the waste initially is hot and radioactive. After immersion in water for some time (about two years for rods from a power reactor) the rods become less hot. Over time, radioactivity deceases. Yet, some of the long-lived elements will be radioactive for thousands of years. Some of the "spent fuel," which really is partly used fuel, can be reused, but that process is costly and dangerous. Hence, the generic plan has been to excavate 1000 feet (304.8 meters) or more below the surface and store the waste underground where emissions, if there were any, would not quickly disperse radioactive elements all over the area. From a risk perspective, underground storage is sensible rather than above-ground storage in grout and metal casks, all other conditions being equal.

An acceptable underground site should have the following properties:

- Stable geological structure;
- No nearby underground or surface water that could compromise the storage area;
- Bedrock with impervious properties;
- Ability to withstand an underground release without dispersing radioactive materials into the atmosphere;
- Ability to monitor the status of the storage site and intervene, if necessary;
- Distance from population centers;
- Access for shipments coming into the site;
- Ability to secure the site; and
- Ability to leave messages for future generations who may speak a language not currently in existence.

WIPP

The Waste Isolation Pilot Plant (WIPP) site has many of these attributes. Located in Eddy County, New Mexico, in the Chihuahauan Desert about 26 miles east of Carlsbad. WIPP it is part of the diamond-shaped area that waste managers have been focusing on for nuclear and highly toxic metals (Figure 7.1). WIPP is the westernmost point and about 90 miles east is Waste Control Specialist, Inc. in Texas, which has been targeted by DOE for storing elemental mercury and is already managing low-level nuclear waste (Chapter 6). The southernmost point about halfway between Andrews and WIPP hosts the National Enrichment Facility, near Eunice New Mexico. The facility uses a gas centrifuge to enrich uranium. The northernmost point, located about 15 miles west of Hobbs New Mexico, hosts International Isotopes, Inc., which recovers depleted uranium hexafluoride and designates the recovered materials for other purposes.

Table 7.2 Principal points of the facility siting credo and additional thoughts

Credo guideline	Additional thoughts
"Institute a broad participatory process"	It is important to be completely transparent with participants at every stage about whether the process might or will include a mediator, will end in arbitration, involves direct negotiations, and other process specifics Participants should have the ability to deal with multiple inputs from different parties and methods, and the leader should have technical support to explain options and results to participants Process should include discussion of role of constraint mapping, economic analysis, and other tools to be used to gather and interpret information
"Achieve agreement that the status quo is unacceptable"	Explain current coping mechanisms and what other options will be considered if a new site is not found Be clear about tradeoffs among health and safety, economic, social and other considerations are considered by the parties in their preferences Discuss how tools such as pollution prevention, constraint mapping, clamping, economic analysis, and letting it go figure into the negotiation about this issue
"Seek consensus"	Choosing a representative advisory group is essential to achieving a consensus composed of people who are trusted and have knowledge, expertise, and are open-minded Charrettes/scenarios can help a group reach a consensus around options Use surveys and focus groups to better understand regional population views and preferences
"Work to develop trust"	Need nimble process, strong and agile leadership with ability to bargain and make choices Explain why some activities required new sites, can be clamped to others, and others would need to be let go Be clear about how the organization will develop and maintain scientific input, communicate with the community on an ongoing basis, and involve parties in monitoring and surveillance, as well as other site activities
"Choose the solution that best addresses the problem"	Explain how each plausible solution addresses the problem, and how the solutions will impact important interested parties and which groups will disproportionately benefit and be impacted Explain how disproportionately impacted groups and areas will be addressed Be able to explain how plausible options can be enhanced through negotiations using charrettes and analytical tools
"Guarantee that stringent safety standards will be met"	Present organization's principles for protecting health, safety and the environment Explain how the organization plans to use best technology and human processes and how it will communicate with the public about these Illustrate with organization's record
"Fully address all negative aspects of facility"	Describe how negative impacts will be addressed over time by what group and with what process Describe how compensation will be used to address negative economic aspects for individuals and the communities, and how other issues will be addressed Regional economic impact models could help negotiators understand who benefits and who might be economically impacted

"Make the community better off"

Negotiate benefits that include payment for services, but also for schools, fire, police, recreation and others

Demonstrate how economic impact analysis, constraint mapping and other tools could help better understand community benefits and costs

Give an example(s) from organization's history to illustrate

"Use contingent agreements"

Offer an agreement about how the developer will monitor health, safety, environmental, economic, labor and other agreements, and how it will respond to new information

Explain how the community will be notified and updated if the facility has to be temporarily or permanently closed, or if the company wants to expand the site. What role the community will play in the process?

Explain how the organization has used contingent agreements in the past

"Seek acceptable sites through a volunteer process"

Be clear about why the organization wants the sites

Establish legitimacy by developing a process for a volunteer to submit a proposal and negotiate an agreement

"Consider a competitive siting process"

Specify how the organization will assess the proposals

Explain how the process will evolve

"Work for geographical fairness"

Describe how the community will be able to propose incentives and benefits

As much as possible be clear about what local attributes will make a site unacceptable

Articulate a principle of geographical fairness

Use constraint mapping to demonstrate that the proposed site will not contribute to social and/or environmental injustice, especially exacerbate them

Illustrate how the organization has followed this process in the past

"Set realistic time tables"

Set realistic schedules within the organization

Give organizations sufficient time to weigh their options

Provide organizations support to reach timely decisions

"Keep multiple options open at all times"

Use constraint mapping, pollution prevention, continuous safety improvements, and other options to create multiple options to consider

Articulate multiple options within the organization for siting and non-siting options

Explain within and outside the organization that it has been developing options and what these are

Note
First column from Kunreuther et al.[19,20] Column two contains my comments.

Figure 7.1 Selected nuclear waste sites.

For those that wish to explore this in more detail, I recommend that you type in Waste Control Specialists in Andrews Texas, find the New Mexico–Texas border and then zoom in on the site. You will see what seem like houses near the nuclear sites and then you find out that they are oil wells, not suburban houses. This is a good example for class in photo interpretation, the CLAMP policy, and the tendency to put noxious facilities on borderlands.

The United States federal government originally chose Lyons, Kansas, for nuclear waste storage because of a large salt dome.[21] That site was not developed because of concern that nearby oil and gas wells would compromise the geological formation, as well as strong public opposition. When some residents and elected officials from the Carlsbad, New Mexico, area expressed interest in hosting the proposed site, the DOE investigated the 600-meter thick Delaware salt formation. Covered by a 300 meters of soil and rock, the federal government concluded that the bedrock is virtually impermeable and parts of it had not been disturbed for millions of years. During the author's most recent visit to the WIPP site, he was presented with a plastic bag containing salt that said the salt was 250 million years old.

The plan was to build storage chambers 2150 feet (600 m) into salt formations. The working expectation within DOE was that any voids or cracks that developed over the years would be filled in by salt bending like plastic into voids. The following list is of events that characterize the early part of the successful WIPP facility siting for transuranic waste:

- 1978: Formation of the New Mexico Environmental Evaluation Group (EEG);

- 1979: U.S. Congress authorizes construction of the facility;
- 1979: Congress redefines waste to be stored at the cap WIPP site, limiting it to transuranic waste, which is classified as lower-level waste (see below).

These initial three steps included a great deal of formal and informal bargaining. The creation of the EEG is evidence of negotiation. EEG was charged with addressing concerns about geology, hydrology, soils and all of the technical issues of an underground repository for New Mexico. The group was directed by scientist Robert Neil. As part of his job, Neil listened to many concerns about health and safety from New Mexico residents. Neil[22] listed the public's initial concerns about the site, and I have paraphrased these:

- The federal government should not use the relatively low risk of radiation exposure as a justification for siting the repository because it is an involuntary siting;
- The scientific community needs substantially more resources than it has to evaluate the risks;
- Many solutions for disposing of radioactive hazardous waste have been discontinued, and the public is concerned that the DOE will abandon the site and its responsibility to the area;
- An insufficient number of technically trained public health scientists are available to manage the risks;
- DOE's rules for shipping and burying the waste appear to be inconsistent.
- Distressing violations in DOE's procedures and enforcement have been found.

Each of these bullets demonstrates that New Mexico's leaders, scientists and public did not trust the DOE's technical and communications capabilities, and it more subtly expresses concern about the DOE's values. Given these kinds of concerns, in 1978 the DOE, despite its primacy agreed to fund the EEG. Then, New Mexico sued the DOE and signed an agreement in 1981, which was amended in 1982. These negotiations were the first major negotiations that I am aware of between the U.S. federal government and a state government about siting a repository. The negotiations focused on generic issues that I have seen in every noxious facility case:

- State liability indemnification should there be an event or alleged exposures;
- State ability to monitor the site environment and transportation of nuclear material, in addition to what the federal government does;
- State role in upgrading and maintaining key highway links;
- State role in emergency preparedness;
- Federal and state government agreement on a formal protocol about health and safety issues that might arise;
- Federal government payments to the state for providing services; and

- Federal government agreement to provide technical reports to the state for review in a timely manner.

The DOE–New Mexico negotiations also addressed several WIPP-specific issues. Two were critical. One was that the DOE was willing to provide funds to investigate brine found in the salt bed and the apparent disturbance in that bed. The research and recommendations of the DOE and EEG led to the relocation of the site within the bed, and increased the credibility of both in the eyes of the public. Had this issue not been negotiated, this site, I believe, would have been abandoned.

Neil[22] described four other negotiated changes in the original DOE idea for the site:

- 1000 spent fuel assemblies were eliminated from the storage plan;
- Experiments with high level nuclear waste were reduced 90%;
- The Nuclear Regulatory Commission (NRC) did not need to license the site; and
- The state could not veto the site.

Clearly, the state got some of the things that it wanted and the DOE got some things it needed from these negotiations. Both improved their credibility in New Mexico.

During the first four years, the New Mexico EEG worked on engineering processes, environmental impact, facility design, geology, facility operations and waste acceptance criteria. Reading through the 19 reports prepared by the EEG, they focused on five issues that led to changes in the following:

- Site selection criteria;
- Transportation accident avoidance and management;
- Water pollution associated with breaches and leaching;
- Assessment of geological features that could compromise the site; and
- Improvement in the predictability of geological stability.

Fast forward to the 1990s, and the site was not yet open. In 1991, a federal judge decided that the U.S. Congress must approve the WIPP site before any waste could be brought to the site. Congress approved it in 1992, and the EPA approved revised safety standards. In 1994, Congress ordered Sandia National Laboratory to evaluate the site using EPA's criteria. The report was completed, and the first nuclear waste arrived from Los Alamos National Laboratory in 1999. From hatching the idea to first delivery took over three decades. That feels like a long time. Yet, I was shocked when the site accepted waste. I had never expected this site to open. I waited for something or someone to stop it before March 26, 1999.

After more than two decades of operation what does the local population think about the WIPP site? Hank Jenkins-Smith et al.[23] studied 35 New

Mexico-wide surveys conducted from 1990 to 2001. Support to open the site increased from 42% to 63%. The authors show that support increased with each success, such as when Congress approved the DOE application, when EPA certified the site, when the site opened and when it received waste in 1999. Some of this support is related to the strong backing from Peter Domineci, former U.S. Senator (Republican–New Mexico) and from local Congressmen. In regard to demographics, consistent with the literature, older, white males with a tendency to vote Republican disproportionately were supporters. Notably, also the strongest supporters were in the host area – 73% approval among those within 10 miles compared to 50% at 200+ miles. This is also consistent with previous studies that found more support near existing sites than further away (see below).

Does this imply anything about expanding WIPP's mission to include high level nuclear waste? In essence, there are five plausible options for dealing with the high level nuclear defense and commercial waste:

1   Continue to store commercial nuclear waste at the power plant that generated the waste.
2   Store defense waste at existing DOE sites where it already sits.
3   Open the Yucca Mountain repository in Nevada for high-level defense and/or commercial waste.
4   Find a new or add to an existing deep underground repository at WIPP; and
5   Find another "temporary" site to store the waste, such as one of DOE's former defense sites (Hanford, Savannah River, Idaho, and Oak Ridge).

In order to store the waste using options 3–5, the waste would need to be converted from a liquid to a solid form, placed in storage casks, and transported. The assumption is that few sites would be willing to acknowledge any interest in storing this waste or having the waste pass by their home on a train or truck. To test this expectation, during the summer of 2013, we surveyed 920 people living within 50-miles of four major DOE defense sites: Hanford, Idaho, Savannah River, and the Waste Isolation Pilot Plant. Based on previous surveys, we expected slightly more interest at the WIPP site, which has been storing nuclear waste.[24]

Methods and results of this study are reported elsewhere in detail.[24] Table 7.3 shows some of the key findings, indicating little support for clamping on this function to their local DOE site. Four percent were strongly in favor of the site, 10% somewhat in favor and 4% neutral. Eighty-two percent were opposed or did not answer. The WIPP site, as expected, demonstrated the most interest and support (24%).

While less than 20% would even consider a site being clamped in their area, their strongest attributes are noteworthy.

• They trust the DOE (not its contractors) to manage new on-site projects;
• Are college-educated;

Table 7.3 Support for storing commercial nuclear spent fuel waste at selected DOE sites

Option (n = 922)	%
Strongly favor	4.1
Somewhat favor	10.4
Neither favor nor oppose	3.9
Somewhat opposed	16.1
Strongly opposed	21.4
Maintain current policy not to move waste	36.4
Don't know or refused	7.7
Total	100.0
Strongly favor, somewhat favor, or neutral to hosting in their region	
All sites	18.4
Hanford	19.1
Idaho	14.3
Savannah River	15.2
WIPP	23.9*

Source: Selected data from Greenberg et al.[24]

Note
*WIPP proportion significantly higher than Idaho and Savannah River at $p < 0.05$.

- Are 18–49 years old;
- Want more information about the idea;
- Are relatively affluent;
- Self-identify as political independents; and
- Live in the WIPP area.

Fast forward again to 2014. Two incidents occurred, one of which some found amusing because it was caused by a staff member using organic cat litter instead of clay-based cat litter. But nothing about it was funny. A drum burst, releasing a small amount of radioactive uranium, americium, and plutonium. Some of the monitoring equipment did not work, very low levels of radiation were emitted, and workers were exposed to low levels. The net result was the site was closed for three years, and the DOE spent $500 million (the original estimate was $2 billion), including upgrading the ventilation system, and training increased.

But these problems were not supposed to occur! These incidents hurt the agency's credibility, at a time when its credibility had rebounded from the Fukushima nuclear power plant problems in Japan, for which there was a general loss of credibility in the DOE and other nuclear agencies.[25]

WIPP reopened on January 9, 2017. In June 2017, Leone[26] reported that an inspector intercepted an improperly sealed container in a shipment from Idaho to WIPP. It may be that the idea of storing high-level waste is less popular than before the year 2014 incidents. Yet, despite the problems of 2014, barring another incident, the site should able to complete its mission and possibly expand it.

Arguably, WIPP despite its problems in 2014 and the reality that today's idea of public participation as articulated by the facility siting credo was not followed has to be rated a success, one of the few in the world involving a repository. It opened, has accepted waste, and after its 2014 failure, WIPP reopened. I find no evidence that the local community and/or state is ready to fight to close it. New Mexico has acted as the public's public health agent, and the community seems to accept that role. I have no doubt that without the negotiations that led to the founding of the EEG and the agreement between the federal government and New Mexico, this site would not have opened.

There will be more tests in this area because others will seek to store additional hazardous materials in the area from western Texas to eastern New Mexico.

A few months after writing the first draft of this chapter, the next test arrived in the form of a proposal to reduce the potency of radioactive plutonium and then store it at Yucca. This will be more than a challenging test. Below are my notes when I visited the WIPP site in early 2018.

Text box 7.1 Visit to WIPP, 2018

How weird. I was supposed to visit the WIPP site in 2014, but then the infamous kitty litter incident stopped that trip and closed the site. I'm not supposed to talk about what I see when I go down into the repository, but there is plenty to see above ground. Starting with flying to this site, there is no convenient airport. I talked with three people, each who recommended a different place to fly to. Now that I'm here, I see what the DOE means by remote. Few people, lots of dry climate scrub vegetation and no direct route to the site. Lots of oil rigs near the site. I see a large truck with three large containers, presumably these are carrying transuranic waste bound for the site. Off the right is the sign "Waste Isolation Pilot Plant." Now we drive to the Credentialing Center, and then it's down into the salt chambers.

Yucca Mountain

If WIPP is the world biggest "success story" at finding a repository for dangerous nuclear waste, then Yucca Mountain by far has been the biggest and most expensive failure to site a facility. The essence of the Yucca Mountain story is that the U.S. federal government has already spent well over $10 billion to build a portion of a repository, and it is unclear whether it will ever open. Much has written about this failure. My focus is on major factors that led to the failure and what, if anything, negotiations might have contributed to a more successful outcome.

I begin with the two most salient points. The Nuclear Waste Policy Act of 1982 indicated that the federal government would search for multiple possible repository sites, at least one east of the Mississippi River and one west of it. In

1984, the DOE announced ten sites in six states. After a year in which the states hired consultants to find reasons not to locate in their state, a report was produced about the 10 sites. President Ronald Reagan chose three:

- Hanford, Washington (DOE nuclear defense site located along the Columbia River in southeast Washington);
- Deaf Smith County, Texas (beef country with a population of about 20,000, located about 200 miles directly north of Andrews County, Texas [see Chapter 6]); and
- Yucca Mountain, Nevada (located on the Nevada Test site between Death Valley and Nellis Air Force Base).

However, Congress had neither the patience nor the willingness to find out if any of these were plausible, and in 1987 it designated Yucca as the site. This infuriated Nevada and led to the often heard expression that the Nuclear Waste Policy Act of 1982 was really the "screw Nevada" act (actually, it was the changes made in 1987 to the law that designated Yucca).

Second, the State of Nevada heavily depends upon tourism, and the state government and the business owners are adamantly opposed to a site roughly 100 miles to the north of Las Vegas, which they are convinced will hurt visitors to Las Vegas. The fact that Yucca Mountain's neighbors are Death Valley and Nellis Air Force base have not convinced the Vegas-centric population that the Yucca Mountain area site is acceptable. The fact that the waste that would go to Yucca is currently stored at the DOE's major nuclear sites and at over 60 nuclear power facilities on the surface is riskier, or moving it to one of the DOE's sites (Hanford, Savannah River, Oak Ridge, or Idaho) for "temporary storage" is not persuasive for Nevada residents. For my relatives and their friends who live in the Las Vegas, the issue is much simpler: the federal government declared war on Nevada, and Nevada is fighting back.

Surveys document public discomfort. A 1998 survey by Kunreuther et al.[27] elicited the views about the repository from 1201 non-Nevada U.S. residents and 1001 Nevadans. Several observations are notable:

- 53% of Nevada residents and 49% of U.S. ones agreed that a repository is the best way to store high-level waste;
- 45% of Nevadans believed the site would stimulate nearby economic growth compared to 28% of U.S. residents;
- 35% of Nevada residents considered the repository a very serious threat (8–10 on a 10-point scale) compared to 35% of their U.S. counterparts.

These answers suggest that residents of Nevada were not different from the U.S. as a whole. And yet only 30% of Nevada residents felt that Nevada is the best place locate the site. The authors queried respondents about their willingness to support the site if they received an annual tax benefit of $1000, $3000, and $5000 a year for 20 years. Only about one-fourth of respondents were willing to trade support for financial benefits, and support was not notably impacted by the dollar benefit. The only factor that appeared to be correlated was risk perception, that is, those who did not perceive much risk were more willing to accept the benefit. These data should not be interpreted as implying that compensation never works. A large set of studies shows that it is typically part of any agreement between developers, government and individuals.[28–31] However, in this case, the angry public rejected compensation; they felt betrayed by the national government.

I visited my uncle and cousin once a year, both residents of Las Vegas; they were angry that I even visited the site twice and I will spare the unprintable language they used to describe the Department of Energy. This text box is from a trip I made to Yucca Mountain during the period October 23–27, 1999.

Text box 7.2 Uncle and cousin's views of Yucca Mountain proposal, 1999

I love visiting my uncle Sol (former U.S. immigration services officer) and cousin Morty (pharmacist). Both moved to Las Vegas, my uncle after spending his entire adult life in New York. Both are avid Yankee fans, and they want the Yankees to beat the Atlanta Braves in the World Series again. But they are ready to strangle me because I'm visiting Yucca Mountain as part of my work for the DOE.

Uncle Sol: we don't want any of that radioactive crap in our state. I know it is 100 miles away, but it could kill the economic expansion of this state. It is bad enough that we got stuck with storing mercury [he means at the Hawthorne Army Depot southeast of Reno, see Chapter 6]. The government [federal] is messing up this state. It's not fair, not tolerable.

Cousin Morty: I agree. I don't have a better solution, but I read some reports on this and learned that Congress arbitrarily picked Nevada. People are really angry and fighting to keep this site from opening. We want a baseball, football or basketball team, not radioactive waste. People here will spend their own money for a long time to stop Yucca from opening. [He is right, they have.]

Michael Greenberg, last thoughts while waiting for a flight back to New Jersey: Well the Yankees swept the Braves. But this site is even more interesting than the World Series. You drive north from Vegas. The terrain quickly turns to brush and desert. Was I imagining things, or does it seem that bullet holes in the road signs have doubled since I left Vegas. The closer I get to Yucca, the more the signs have bullet holes in them. Is that a message to the feds? ... Jets periodically fly overhead, I guess from Nellis Air Force Base.... The tunnel [inside Yucca], that's it? It is a mini-version of New York City and DC. A lot of money spent, but no passengers. Death Valley was a lot more interesting. Actually, the tunnel is not interesting.

Until the election of Donald Trump as President of the United States, Nevada was protected against Yucca. Democrat Harry Reid, Senate Majority leader from Nevada, working with President Obama, took the site off the table. Indeed, DOE secretary Chu stated in 2009 that the site was no longer being considered, and a year later the DOE moved to withdraw its application made to the Nuclear Regulatory Commission.[32] However, others have contended that this was a political decision, not a science or engineering based decision and several states with stored commercial nuclear waste sued the DOE to move the project forward and to refund money being gathered from taxpayers and utilities to fund the site's development.[33-40]

During the period 2004–2014, the opposing parties have been in court with both sides wining skirmishes. The charges and counterstatements further highlight the basic fight over science, trust and especially political legitimacy,

Now, in 2017, there are strong rumblings that the new Republican-led government will try to reopen siting Yucca. Nevada, assuming that this will be tried, is already assembling a war chest to fight the idea.[41] Might they succeed? Pflum[42] examined the institutional environment of the Yucca and WIPP. He concluded that Yucca has stronger laws, better regulations, and a more experienced and balanced oversight. The only advantage he cedes to New Mexico is its state government. If so, why hasn't Yucca opened, and why do I believe that it will not open unless the current party in power is willing to risk the political wrath of a state. Even if Yucca is a better choice from the perspectives of human health and safety than leaving the waste at the existing sites or moving it to "interim" storage site, the fight is about power and legitimacy not science, although a lot of the words are about science.

The two issues I noted above – U.S. government behavior toward Nevada, and the Nevada tourism argument leads me to the conclusion that Yucca is likely to remain a six-mile long tunnel from nowhere to nowhere. The state of Nevada was treated in a most disrespectful way by the U.S. Congress, and I see no near-term expectation that they suddenly will change their mind and begin to trust the federal government. In contrast, New Mexico sued and negotiated a changed venue and resources to form an expert group of its own. The Nuclear Waste Technical Review Board (federal) may have stronger science and engineering than the Environmental Evaluation Group (New Mexico) but the EEG certainly has been effective at getting the attention of the federal government. The Environment Protection Agency and the Nuclear Regulatory Commission have strong science, but if I wanted to work with a group that would be more willing to listen to my views across a spectrum of science and social science issues, I would choose the EPA and its broad expertise. Contrary to Pflum, I think New Mexico wisely negotiated institutional arrangements for itself.

I do not understand how Congress expected any other outcome when it chose Yucca Mountain and threw sand in Nevada's face. In contrast, the southeast corner of New Mexico clearly has sent signals about wanting these kinds of sites. Southeast Mexico has had a need for new economic base to replace its unpredictable oil and potash industries.

In short, WIPP and Yucca would be the United States' candidates for high stakes, politically costly gambles that federal legitimacy would be sufficient to site a noxious facility. It is possible that WIPP may yet close, especially if there is another serious incident. But it may also be the place that is willing to accept high level nuclear waste and other nuclear materials. Yucca may open at some point, but for what functions are something that I can only speculate about.

International examples

There are remarkable negotiation examples about siting geological disposal facilities for high level nuclear wastes from other countries. Finland is planning to be the first country to open a permanent repository by 2023, which is estimated to cost about $3.9 billion.[43,44] According to the 1987 Nuclear Energy Act, companies running nuclear facilities are responsible for managing their waste and all related costs. In addition, the Ministry of Economic Affairs and Employment of Finland has an approval role, and the Radiation and Nuclear Safety Authority of Finland (STUK) monitors the implementation of disposal programs.[43] Selection of the site is mainly based on a volunteer process, and host communities have veto power.[45] After initial research led to six possible candidate locations, the municipality of Eurajoki was proposed to host a permanent disposal facility in 1999.[46] Successful local engagement stands out throughout the negotiation process of Onkalo Repository.

Sweden is considered one of the most advanced countries in terms of negotiating with potential host communities for disposal facilities. The Stipulation Act of 1977 shifted responsibility of radioactive wastes and all related costs from the government to the industry, and Swedish Nuclear Fuel and Waste Management Company (SKB) was founded.[47] In addition to partnership agreements with SKB, host communities also have veto rights.[45] After 30 years of research, Östhammar is considered to be the best option to host a permanent nuclear waste disposal facility. The fact that 88% of Östhammar residents approve the project shows the success of the negotiating efforts from the beginning.[48] SKB carries out legally administered negotiations with local government, communicates with the community, interacts with local leaders, and even organizes guided tours of the facilities.[49]

Disposal of high-level nuclear wastes is a controversial issue in Germany. Until recently, Gorleben was considered one of the options to host a repository for high-level radioactive waste. However, this small municipality has also become a symbol for anti-nuclear protests.[50] Controversies regarding site selection and constant opposition from activists and politicians led to a search for alternative options for a permanent repository.[51] The Federal Office for Radiation Protection of Germany is responsible for disposal of nuclear wastes. Their website highlights transparency and public participation in the site selection procedure. Germany is planning to conclude site decision research by 2031.[52]

It will not surprise you to note that the most likely successful siting opportunities in these countries are places that already have a major nuclear facility in the area, in other words, what is also the case near the WIPP site in the United States.

Apart from the geological disposal of high level nuclear wastes, there are other interesting negotiation examples from other countries on siting noxious facilities to track. For instance, Taiwan has been promoting incinerators for waste disposal and using different compensation strategies in order to persuade community-based oppositions. Nevertheless, limited opportunities for public participation is one of the main reasons for ongoing local opposition.[53] Separately, in addition to public health concerns, religious sentiments may also influence decision-makers' positions on noxious facilities. For instance, the sacredness of the cow in Hindu India is an important factor for locations of slaughterhouses, which have recently faced restrictions.[54]

Final thoughts

Negotiations are a way of trying to avoid head on collisions between rival ideas of justice that are so apparent in many noxious facility siting cases. It is tempting not to negotiate when you have more power and resources than your opponents or at least think that you do, and when you are certain that your choice is technically superior to other options, cost effective and, and morally acceptable, at least as you define morality. Nevertheless, decision-makers must fight these feelings in the case of siting noxious facilities and always be prepared to negotiate, and to call for engagement once key players have been identified.

WIPP and Yucca Mountain are two American icon noxious facility sites located about 1100 miles apart, but a world apart when it comes to politics, economy, the success of negotiations, and siting nuclear waste repositories. WIPP is hardly the model of broad participation that today's practitioners call for. But it opened and has reopened after a problem, and the state and community have not pressed for it to be closed. Yucca has a six-mile long tunnel that costs billions of dollars, and spent nuclear fuel and defense nuclear wastes sit at existing DOE and nuclear plant locations.

Would the results have been different if the federal government had begun with negotiations in the core of the siting process, using something like the facility siting credo, or the six guidelines I suggest in Chapter 10? The searches would have had at least the following four steps:

- Involve local government in deciding if it wants to consider a site;
- Conduct a feasibility study to see if the sites meet environmental, social and other siting criteria;
- Determine if any of the communities that pass the first two steps want to compete for the site;
- Engage in economic, social and political negotiations to select a final site.

Neither of the two iconic U.S. sites was in the mind of the federal government at first. WIPP eventually had all of them, but not until the Congressionally mandated choices failed and the DOE had to turn to a negotiations-centered model. The brief international section considers the cases of Finland, Sweden, and Germany. Finland and Sweden have been praised for following negotiation-oriented processes. The above negotiations-heavy process gets us away from one side arguing that the other is greedy, uncaring, and threatening pre-emption and the other from countercharging that the opponents are parochial, inflexible, irresponsible, and self-serving. But I am not thinking of this model as a panacea, even though it has more legitimacy in a democratic context. I am not persuaded that it will work because some sites are acceptable locally and/or physically unfit. In the gray world of risk analysis, no one model is going to work.

References

1 Brown H, Marriott A. *ADR Principles and Practice*. 3rd ed., London, London Sweet & Maxwell/Thomson Reuters, 2011.
2 American Arbitration Association. https://ww.adr.org.
3 Fisher R, Ury W. *Getting to Yes: Negotiating Agreement Without Giving In*. New York, Penguin, 1981.
4 Forester J. *The Deliberative Practitioner: Encouraging Participatory Processes*, Cambridge, MA, MIT Press, 1999.
5 Tenner E. *Why Things Bite Back: Technology and the Revenge of the Unintended Consequences*. New York, Alfred Knopf, 1996.
6 Kahneman D. *Thinking, Fast and Slow*. New York, Farrar, Straus and Giroux, 2011.
7 Clark G, A Theory of Local Autonomy. *Annals of the Association of American Geographers*. 74, 195–208, 1984.
8 Andrews R. Hazardous Waste Facility Siting. 117–128 in C Davis and J Lester, eds. *Dimensions of Hazardous Waste Politics and Policy*, New York, Greenwood Press, 1988.
9 Lake R, Johns R. Legitimization Conflicts: The Politics of Hazardous Waste Siting Law. *Urban Geography*. 11, 5, 488–508, 1990.
10 Boholm A, Lofstedt R. *Facility Siting: Risk, Power and Identity in Land Use Planning*. New York, Taylor & Francis/Earthscan 2013.
11 Morrill R. Inequalities of Power, Costs and Benefits across Geographic Scales: The Future Uses of the Hanford Reservation. *Political Geography*. 18, 1–23, 1999.
12 Greenberg M, Amer S. Self-interest and Direct Legislation: Public Support of a Hazardous Waste Bond Issue in New Jersey, *Political Geography Quarterly*, 8, 67–78, 1989.
13 Berzok J. Toxic Siting Panel Sued by Bedminster Group. *Star Ledger* (Newark), April 8, 1, 14, 1986.
14 Johnson T. Potential Sites Listed for Waste Facilities. *Star Ledger* (Newark), February 15, 1, 11, 1986a.
15 Johnson T. Problems Snag State's Selection of Waste Sites. *Star Ledger* (Newark), April 25, 36, 1986b.
16 New Jersey Hazardous Waste Facilities Siting Commission. *New Jersey Hazardous Waste Facilities Plan*. Trenton, NJ, Hazardous Waste Facilities Siting Commission, 1984.
17 Dodd F. Siting Hazardous Waste Facilities in New Jersey: Keeping the Debate Open. *Seton Hall Legislative Journal*. 9, 2, 423–436, 1986.
18 Johnson T. Potential Sites Listed for Waste Facilities. *Star Ledger* (Newark), February 15, 1, 11, 1986a.
19 Kunreuther H, Susskind L, Aarts T. *The Facility Siting Credo: Guidelines for an Effective Facility Siting Process*. Philadelphia, PA, University of Pennsylvania, 1996.
20 Kunreuther H, Fitzgerald K, Aarts T. Siting Noxious Facilities: A Test of the Facility Siting Credo. *Risk Analysis*. 13, 3, 301–318, 1993.
21 Walker S. An "Atomic Garbage Dump" for Kansas. *Kansas History*. www.kshs.org/pubicat/history/2006winter_walker.pdf. Accessed June 16, 2017.
22 Neil R. New Mexico Environmental Evaluation Group Experience in Reviewing WIPP. www.wmsym.org. (EEG). Accessed June 3, 2017.
23 Jenkins-Smith H, Silva C, Nowlin M, deLozier C. Reversing Nuclear Opposition: Evolving Public Acceptance of a Permanent Nuclear Waste Disposal Facility. *Risk Analysis*. 31, 4, 629–644, 2011.

24 Greenberg M, Mayer H, Powers C, Kosson D. Nuclear Waste Management and Nuclear Power: A Tale of Two Essential United States Department of Energy Sites in Idaho and New Mexico, in *Nuclear Portraits*, L. MacDowell, Toronto, University of Toronto Press, 2017, 217–237.
25 Greenberg M, Weiner M, Kosson D, Powers C. Trust in the U.S. Department of Energy: a Post-Fukushima Rebound. *Energy Research & Social Science*, 2, 145–147, 2014.
26 Leone D. Improperly Sealed TRU Waste from Idaho Shipped to WIPP. *Weapons Complex Monitor*. 28, 22, article 4, 2017.
27 Kunreuther H, Desvousges W, Slovic P. Nevada's Predicament. *Environment*. 30, 8, 16–20, 30–33.
28 McMahon R, Ernst C, Miyares R, Haymore. Office of Solid Waste, U.S. EPA. *Using Compensation and Incentives When Siting Hazardous Waste Management Facilities: A Handbook*. SW-942. Washington, DC, U.S. EPA, 1982.
29 Schmeidler E. *Negotiation Over Hazardous Substance Issues: A Selected, Annotated Bibliography of Case Studies*. Washington, DC, Council of Planning Librarians, CPL Bibliography 277, 1991.
30 Schmeidler E. *Research on Mitigation, Compensation, and Incentives in Hazardous Waste Facility Siting: A Partially Annotated Bibliography, P2432*. Monticello, IL, Vance Bibliographies, 1988.
31 Boyle S. *Siting New Hazardous Waste Management Facilities through a Compensation and Incentives Approach: a Bibliography*. Chicago, IL, CPL Bibliographies. No. 122, 1983.
32 Josef H. Nuclear Waste Won't Be Going to Nevada's Yucca Mountain. Obama Official Says. *Chicago Tribune*. March 6, 2009, 4, 1.
33 Environmental News Service. Nevada Sues Again to Block Yucca Mountain Nuclear Repository. September 10, 2004. www.ens-newswire.com/ens/sep2004/ 2004–09–10–03/ asp. Accessed July 22, 2014.
34 Lindberg M. Yucca Mountain Nuclear Repository One Step Closer to Licensing. September 9, 2008. http://planetsave.com/2008/09/09/yucca-mountain-nuclear-repository-one-step-closer-to-licensing/. Accessed March 27, 2018.
35 Wald M. Is Yucca Mountain Still Dead? *New York Times*. May 3, 2012. https://green.blogs.nytimes.com/2012/05/03/is-yucca-mountain-still-dead-2/. Accessed March 27, 2018.
36 Northwest News Services. Federal Judge: Yucca Mountain Licensing Must Move Forward. August 15, 2013. http://nwnewsnetwork.org/post/federal-judge-yucca-mountain-licensing-must-move-forward. Accessed March 27, 2018.
37 Yucca Mountain Organization. What's New – Yucca Mountain in the News. July, 2014. www.yuccamountain.org/new.htm. Access July 22, 2014.
38 Sebelius S. Watch Your Step Around Yucca, Adam Laxalt, Lest you Fall In! *Las Vegas Review Journal*. www.reviewjournal.com/uncategorized/watch-your-step-around-yucca-adam-laxalt-lest-you-fall-in/. Accessed March 27, 2018.
39 Editor, Environment News Service. Court Forces Yucca Mountain Decision on U.S. Nuclear Regulatory Commission. August 13, 2103 http://ens-newswire.com/2013/08/13/court-forces-yucca-mtn-decision-on-u-s-nuclear-regulators/. Accessed March 27, 2018.
40 Northey H. GAO: Death of Yucca Mountain Caused by Political Maneuvering. *New York Times*, May 10, 2011. https://archive.nytimes.com/www.nytimes.com/gwire/2011/05/10/10greenwire-gao-death-of-yucca-mountain-caused-by-politica-36298.html. Accessed March 27, 2018.
41 Leone D. Nevada Lawmakers Approve $7M+ for Yucca Fight From 2017–19. www.exchangemonitor.com/publication/wcm/nevada-lawmaker. Accessed June 12, 2017
42 Pflum C. A Comparison of Two Potential Repositories: the Waste Isolation Pilot Plant and Yucca Mountain. www.osti.gov/servlets/purl/67261. Accessed March 27, 2018.

43 Ministry of Economic Affairs and Employment of Finland. Licensees responsible for nuclear waste management. http://tem.fi/en/nuclear-energy/nuclear-waste-management. Accessed June 23, 2017.

44 Fountain H. Finns Work to Entomb the Waste of Reactors. *New York Times: Science-Times.* June 13, 2017.

45 Di Nucci M, Isidoro Losada A, Brunnengraeber A, 42–48 in S. Fanghaenel, ed. *Same, Same But Different A Comparative Perspective on Participation and Acceptance in Siting Procedures for HLW Repositories in France, Sweden and Finland.* Germany, KIT Scientific Publishing, 2015.

46 Radiation and Nuclear Safety Authority. Disposal of Spent Fuel in Finland. Radiation and Nuclear Safety Authority. 2016. www.stuk.fi/web/en/topics/nuclear-waste/disposal-of-spent-fuel-in-finland. Accessed June 23, 2017.

47 World Nuclear Association. Nuclear Power in Sweden. 2017. www.world-nuclear.org/information-library/country-profiles/countries-o-s/sweden.aspx. Accessed June 26, 2017.

48 Nyberg P. The Town That Wants Nuclear Waste. CNN. 2011. www.cnn.com/2011/WORLD/europe/04/24/sweden.nuclear.waste/index.html. Accessed June 26, 2017.

49 NPR. In Sweden, a Tempered Approach to Nuclear Waste. *NPR.* 2011. www.npr.org/2011/07/28/138707842/in-sweden-a-tempered-approach-to-nuclear-waste. Accessed June 26, 2017.

50 Thurau J. Germany to Dump Nuclear Waste for Good – But Where? *Deutsche Welle.* 2016. www.dw.com/en/germany-to-dump-nuclear-waste-for-good-but-where/a-19380548. Accessed June 23, 2017.

51 Spiegel Online. Germany's Endless Search for a Nuclear Waste Dump. 2010. www.spiegel.de/international/germany/the-curse-of-gorleben-germany-s-endless-search-for-a-nuclear-waste-dump-a-672147.html. Accessed June 23, 2017.

52 Bundesamt für Strahlenschutz. The Site Selection Procedure. www.bfs.de/EN/topics/nwm/repositories/site-selection/site-selection-act/site-selection-procedure.html. Accessed June 23, 2017.

53 Hsu S.-H. NIMBY Opposition and Solid Waste Incinerator Siting in Democratizing Taiwan. *The Social Science Journal.* 43, 3, 453–459, 2006.

54 Al Jazeera Media Network. India Crackdown on Slaughterhouses Stirs Muslim Unease. *Al Jazeera Media Network.* 2017. www.aljazeera.com/news/2017/03/india-crackdown-slaughterhouses-stirs-muslim-unease-170329131203603.html. Accessed June 23, 2017.

8 Letting-it-go

Introduction

Chapter 8 has three objectives:

- Define a let-it-go policy and explain why it occurs;
- Describe and briefly illustrate different let-it-go policies;
- Illustrate a complex let-it-go policy with the metal lead.

A let-it-go policy

I define a let-it-go policy as a change in the economic, political and social environment of an organization that markedly changes its need for certain numbers, sizes, and locations of sites, ultimately leading to closure of sites and even abandonment of manufacturing of the product. Any of the factors described in Figure 4.1 could be triggers, such as the following factors:

- Change in organization values and long-term objectives because of a new upper management, merger, returning to core functions, in other words, a major internal environmental change;
- Change in production and product delivery factors such as raw materials, labor costs, energy, infrastructure, and transportation;
- Changing the value of national currency relative to others, which changes competitiveness;
- Rising demand for products in other countries, which opens new markets compared to stable or even declining markets in the home country;
- Technological innovations, including robotics and the Internet;
- Preferences of upper management and key workers; and
- Legal mandates, including health and the environment, that make some locations more attractive than others.

The advantages and disadvantages of a generic let-it-go policy for noxious facilities are summarized in Table 8.1.

Major advantages of a let-it-go policy are that an organization can rid itself of an activity and sites that it no longer wants in its portfolio, and then can use

Table 8.1 Generic advantages and disadvantages of a let-it-go policy

Criteria	Advantage	Disadvantage
Raw materials and energy	Elimination or reduction of need at site being abandoned, and ability to shift resources to other uses	Resources needed to close site (demolition, decommission, remediation) and for maintenance
Labor	Local labor force no longer needed, shifted elsewhere, or changed to accommodate new uses	Buyout packages, relocation of some workers; labor needed for closing and maintaining site until owners are no longer responsible
Markets	Abandonment of market that is no longer profitable	Loss of some longstanding customers that may have negative ramifications
Capital	Capital available for other projects	Capital needed to pay for site remediation and preparation for sale or reuse
Land	Land available to be sold, for reuse, or for open space	Change in site status will draw attention of government agencies and site inspectors will look for issues that need remediation
Local cooperation	Local government may be pleased to rid area of an eyesore	Local government may be outraged by organization and call for an investigation and resource investments to accommodate local adjustment problems
Infrastructure, including transportation	Avoid the need to add new infrastructure and fully repair existing infrastructure	Sunken investment abandoned, may need to be removed or maintained to avoid safety, health and environmental hazards
Agglomeration	Agglomeration-related investments can be reduced, leaving resources for other investments	New agglomeration economies need to be built if site or other nearby site needs are to be accommodated
Environmental science and technology	Opportunity to arrive at a final solution for at a problematic site	Costly remediation, stabilization could result
Public concerns and environmental justice	Local population may be pleased to see noxious eyesore closed	Loss of jobs and taxes will lead to pressure for a costly high level remediation to avoid long-term negative legacy
Local government autonomy	Cooperation, especially if new jobs replace lost ones and new jobs improve or at least stabilize the local economy	Formal and informal actions to prevent closing by demanding remediation to a residential level and posting of long-term financial guarantees to fix environmental legacy problems

recovered resources to invest in activities and places that are consistent with its image of a twenty-first century portfolio. Major disadvantages are that these transitions come with social, political and economic costs. These include being held to high cleanup standards at the sites they are leaving, and being held responsible for deliberately abandoning their long-term workforce, including reducing or eliminating employee benefits, and the town that hosted them. Likely the organization will be charged with endangering places where the noxious industry will relocate, its people, environments, and cultures with their processes and products – all in order to serve a short-term bottom line.

Evidence of a let-it-go policy

Included in my list of let-it-go options are sending material outside the Earth's atmosphere, off shore to the oceans, and to other countries. The first option is much more for science fiction writers than for those who live in the present world, the water option has had limited success, and the third has been practiced for many years, and I believe will be increasingly difficult to implement.

Send it to outer space

The movie *Soldier* (1998),[1] starring Kurt Russell, tells the story of a highly skilled soldier in the year 2035. The soldier is considered obsolete by his commanders. After defeating a presumably superior genetically engineered soldier, Russell, assumed dead, is taken on a garbage scow that flies into space and dumps its waste, including the barely alive soldier, on Arcadia 234, a waste disposal planet. Russell meets up with people who had crash-landed on the planet, eventually kills the steroid-hyped genetically engineered soldiers, saves many of the stranded people, and brings them to their original destination. Russell's muscles and personality are commanding, and while the movie was not a financial success, it does a good job of showing what a planet for garbage might look like – actually a grand version of some of the landfills I have visited, including several mentioned in Chapters 3 and 10.

Popular science magazines, bloggers and, periodically, genuine scientists have advocated that nuclear waste be dropped into volcanoes and the deepest canyons of the oceans, and lowered into Antarctica's ice sheets.[2] By far, the most popular option is sending nuclear waste into outer space. Jonathan Coopersmith presents three reasons to send nuclear waste into outer space:

- It is safe to send it into space.
- Space is a better disposal site than underground geological formations.
- The effort would help stimulate other uses of outer space.

After concluding that the space shuttle and conventional rockets are not up to the task, Coopersmith, an advocate of more use of outer space, explores other ideas, such as a ground-launched system like a space gun that would use lasers, microwave, and electromagnetic technologies to fire the payload into orbit and orbit it around Venus, for example. In the future, he argues, that valuable radio-active material can be recovered.

Opinion about putting garbage in space ranges from mockery to strong advocacy. I have read blog sites, and the Quora one offers a wide range of views. Many respondents presented engineering and physics as their credentials. I summarize four (not using their identifiers) from a set that was published in 2012 with the title "Why Don't We Dispose of Nuclear Waste into Outer Space?"[3]

- Commenter 1 – "very, very, very expensive," … "even worse, rockets still explode sometimes during takeoff."
- Commenter 2 – "A failure would dump thousands of kg of highly radio-active waste into the environment."
- Commenter 3 – "Too costly, too inherently dangerous."
- Commenter 4 – "Very stupid idea" – "not uncommon for rockets to blow up."

NASA[4] has produced several reports on the option. In 1978, for example, it issued a 118-page report that it would be possible to use a space shuttle in the 1990s to send some high-level nuclear waste into space, but only a small fraction could go to the moon or to a solar orbit. The report identified three problems:

- High launch costs;
- Accidents at launch or while going into orbit require a rescue capability that does not exist;
- Environmental impacts require protecting employees from the hazardous waste, as well as dealing with impacts associated with building, launching, and operating the equipment.

I end this brief section by referring to a year 2016 report by the World Nuclear Association,[5] which posted a list of ten myths and realities about radioactive wastes. Number 9 is that the waste should be disposed of in space. The text states that the option has been repeatedly investigated since the 1970s, it has not been implemented, and there are no new studies. The two major constraints are high cost and potential launch failures.

In the movie Superman IV, the Man of Steel picks up all the nuclear weapons, places them in a giant net, and throws them into the Sun.[6] If such a being were to arrive on Earth, I would hope that s/he would not forget the massive amount of waste as part of the exercise because at the present time it seems infeasible, dangerous, too expensive to let it go into space and it is a challenge to place it on the surface or underground.

The oceans

Water bodies have been an obvious place to dispose of waste with varying degrees of treatment. Here we examine two of the most controversial attempts of the late twentieth century.

Siting nuclear power plants

Writing in the Sunday *New York Times* in 1972 about nuclear power plants, physicist Ralph Lapp.[7] ended his article with an assertion that the U.S. would need at least 1000 reactors of 1000 megawatt capacity by 1990 (see also Chapter 6). He did not think that the heat load associated with these plants could be supported by the U.S.'s rivers. Hence he concluded that "it does not take much imagination" (p. 90) to envision a set of artificial islands along the American east and west coasts almost from border to border hosting nuclear power plants. He used New Jersey as his example of a state that was densely populated, had little space for new power generation, and was unwilling to tolerate the ecological impact of such plants. Lapp then focused on a proposal to build two new nuclear generating plants at a cost of about $1 billion in 45 feet of water less than three miles offshore and northwest of Atlantic City, New Jersey. Lapp then discussed other locations along the Atlantic coast.

The facility itself would be surrounded on the oceanside by a massive breakwater to protect the site against a hurricane. Inside the breakwater would be a 17-story containment building where the reactor, steam system, and various other essential components, such as living quarters, would be placed. The shore-facing side would also have a breakwater. Transmission lines buried in the ocean floor would run underneath the land-facing breakwater to the shore. Safety measures, including beacons would warn boats away from the area.

A little over a year later, this project was supported by a report from the engineering firm Dames & Moore,[8] which had a specialty in siting and designing nuclear power plants. For context, the company considered ideas for nuclear power plants from countries all over the world. I was sometimes involved as a consultant. I say this only to convey a sense to the readers that in the late 1960s and early 1970s, no site was automatically precluded. Proposals came in from countries with active volcanoes near proposed sites, and I recall a group of us indicating that we would not support the idea of sites near active volcanoes.

Compared to some others I saw, the Atlantic Generating Station (AGS) idea was relatively benign. Dames and Moore's report on the AGS contained the following:

- Siting considerations;
- How the location impacts the design;
- Site-specific investigations;
- Siting transmission lines; and
- Other environmental studies required for the project.

The authors methodically went through the kind of work they would need to do in order to locate the proposed Atlantic Generating Station. They stated that the facility had to satisfy economic, environmental, design conditions and be close enough to markets. The advantages were that a site would be easier to acquire than if it were on land, cooling water would surround the site, and the plant would be built elsewhere (Florida) and then brought north, which would eliminate weather as a problem. The conclusion was that offshore nuclear plants had become realistic options.

I was involved with several projects that Dames & Moore undertook, primarily the Newbold Island project in the Delaware River, which was not built. After our population and land use analyses were completed, I testified that locating the proposed site between Philadelphia and Trenton New Jersey could potentially require that over 10 million people be evacuated in the case of a serious event. We realized that the population size was too large for evacuation. By comparison, the Atlantic Generating Station proposal looked better because there was no residential human population to the east of the site, only the Atlantic Ocean.

Like the Newbold Island proposal, the Atlantic Generating Station one failed. The company that was to build the facility in Jacksonville, Florida, had other clients waiting, but the idea was abandoned. President Jimmy Carter came into office in 1976 calling for a moratorium on nuclear power plants. The Three Mile Island Nuclear Power Plant event occurred in 1979, and this combination in essence ended nuclear power plant momentum in the United States. My notes state that the 38-story high crane built to construct the nuclear plants was sold to a Chinese shipbuilding company. Fast forward a quarter of a century, Novak[9] writing for the Smithsonian in 2013 labeled the Atlantic Generating Station and similar plans as "ill-advised," which is perfect 20–20 hindsight that I find disconcerting.

Yet, in fact, the idea has been reborn in Russia, the United States, and China. Russian efforts date from the 1990s, as part of their effort to supply electricity to isolated small cities in the north. They are building several 35 megawatt reactors that can be floated on 150 meter by 30 meter icebreakers.[10] In fact, the U.S. had a small ship that used nuclear reactors to supply electricity to the Panama Canal area, and warships have been powered by nuclear fuel for many years. Of course, these are not the 1000–1100 megawatt reactors envisioned for the Atlantic Generating Station.

China is moving forward with a plan to build 100 and 200 megawatt reactors that would sit on reefs in the South China Sea.[10,11] The power would be used for offshore oil and gas exploration, desalinization facilities and to support the artificial islands China is building in the contested area. Will this plan materialize into 20 larger reactors along the South China Sea? The idea has to be taken seriously because China's resource needs have markedly expanded, including the construction of massive tunnels to move fresh water to the north and expand its use of coal, if another source of electricity is not developed.

In 2014, *The Economist*[12–14] reported that MIT engineers are proposing technologies that would build nuclear power plants along the eastern U.S. seaboard and float them eight miles out to sea. While the author quotes engineers stating that this plan could be a "potential game changer," I am skeptical. The current reality is that nuclear power plants are closing in various locations across the United States because of competition from low cost shale-gas, including locations where the MIT engineers expect nuclear plants to be located. For this second round of considering nuclear plants in the ocean to succeed in the United States, the lucrative shale-gas supply would need to be more expensive. The off-shore nuclear power plant idea might return, but now operating nuclear plants are being closed down and are requesting government subsidies to keep operating.

Incinerating hazardous waste at sea

Incinerators have a long history, and many who live near them have strong negative perceptions. For example, the author's grandmother Rose lived in an apartment house with a building incinerator and nearby to a New York City incinerator. When I mentioned to her that the government was working on a plan to incinerate hazardous waste in the ocean, her response was the Yiddish word "schmutz," which means dirt, stain or filth. When I explained to her that they were going to incinerate liquids, her response was to not believe that statement and she labeled the company as a bunch of "ganefs" (thieves, or swindlers). She, even more skeptical than me, figured they would take the waste out into the ocean and open the valve. Over 15 years later, I heard similar assertions of dirt and mistrust accompanied by much less endearing words from a young woman in Alabama about the U.S. Army's incineration of its chemical weapons stockpile.[15]

This section focuses on burning hazardous waste, not household garbage, albeit some household garbage contains hazardous materials. In 1972, the International Convention on the Prevention of Marine Pollution by Dumping at Sea, more commonly known at the London convention prohibited the common practice of dumping wastes into the oceans. In the U.S. the Marine Protection, Research, and Sanctuaries Act (also known as the ocean dumping act) (PL 92–532) added pressure to develop less polluting waste management technologies in the United States. In Europe, ocean incineration was considered a possibility and the first incinerators were developed and tested there.

With these legal acts as context, incineration at sea became a major issue in the United States during the early 1980s and 1990s when U.S. law, most specifically, the Resource Conservation and Recovery Act (RCRA), made it much more difficult for companies to dispose of waste in landfills. Many gradually turned to pollution prevention (Chapter 5). But their choices at the time were more expensive landfills options, deep-well injection, and land-based or seabased incinerations.

With the exception of some waste management companies that saw a business opportunity, neither land or sea incineration options were welcomed with enthusiasm. For example, writing for Greenpeace, Costner and Thornton's[16]

Playing with Fire: Hazardous Waste Incineration described incineration in such negative terms that leaking landfills seemed more desirable. Incinerators, they argued, do not properly manage heavy metals, dioxin, and other toxic chemicals; incinerators routinely experience equipment failures, human errors; and rapid changes in the waste fed into incinerators are not well managed. The authors recommended a ten-year moratorium on new incineration sites and permits for waste-burring facilities, for specially built incinerators, cement and aggregate kilns, and other industrial facilities that burn hazardous waste. Second, they called for ending the burning of wastes containing chlorine, other halogens, and metals. This was the strongest anti land incineration presentation I have read, but hardly unique (see Chapter 7 for more about siting for hazardous waste incineration facilities on land).

Almost two decades later, Greenpeace[17] had not changed its message. A Greenpeace New Zealand group called for a ban on incineration. The only difference from the earlier report is that they placed greater emphasis on their belief that burning today's waste is too costly and too complex to safely manage than they had in their earlier anti-incineration report.

Nevertheless, an incineration option for hazardous materials, including ocean-based. was evaluated and tested because of government legislation that made new landfills solutions problematic. In Europe tests began in the early 1970s. A cargo ship built in Hamburg was sold to Ocean Combustion Services in Rotterdam. Flying under the flag of Singapore, it incinerated waste in the North Sea. The owner had the ship rebuilt to incinerate chemical wastes, which required adding tanks to hold the waste and two incinerators. The ship, called the *Vulcanus*, had a capacity of roughly 100,000 metric tons a year. It and others like it, began burning some dangerous liquid hazards, including dioxin, left over Agent Orange from the Vietnam War, and other chlorinated hydrocarbons. In 1988, the *Vulcanus* was purchased by Waste Management, Inc., renamed the *Vulcanus I*, and a *Vulcanus II* was added for use in the United States market in response to the federal laws that made landfills options extremely costly.

What occurred during the mid-1980s is a fascinating exercise in environmental management of controversial siting options.[18–21] Writing in the influential journal *Environmental Science and Technology* in 1984, Bond reviewed the evidence for allowing incineration at sea. The EPA issued permits for *Vulcanus* test burns. In 1974, the *Vulcanus* burned organochlorine wastes in the Gulf of Mexico. Other test burns were allowed of Agent Orange and PCB wastes in 1977, 1981 and 1983. Bond reviewed the test burn data and concluded that none of test burns had a sufficient quantity or quality of data to justify a permit. The penultimate paragraph of the paper notes that the *Vulcanus*'s hardware was inadequate for destroying these chemical wastes. Even more to the point, the author argues that proponents of at sea incineration took the faulty data from these trial burns and used it to justify ocean-based incineration.

A notably different view of at sea incineration was presented in 1987 at a conference that this author attended. John Sandstedt,[21] the Director of Environmental Compliance of At-Sea Incineration (ASI) explained the demise of his company. In the end, he argued that the company could not move forward because of opposition by land-based waste disposal businesses, primarily land-based incineration providers, and the unwillingness of EPA officials to act under the pressure of opposition from local officials. ASI had obtained a liability insurance policy, a port facility in Newark, New Jersey, and a lease for land. It conducted several public surveys that Sandstedt asserts showed a clear public plurality for their project compared to landfills, land-based incineration, and conventional waste treatment (I did not see the data).

Sandstedt asserted that their effort was hindered by EPA internal jurisdiction issues, by new EPA staff that had to learn what the existing staff knew (a change in federal administration had occurred), and by efforts of what he called the "land-based incineration lobby" to delay ASI's project. In the end, time was the enemy of the project. His lessons learned are that innovators in this field have to have sufficient capital to wait out interminable delays for permits, have the capacity to front money while waiting, fend off organized opposition groups that use scare tactics, and deal with changing local government leadership.

Is there a future for ocean-based incineration? Incinerators, especially the new ones, are fascinating machines. Different types are used for solids and liquids. Destruction efficiency depends on the combustion temperature, the length of time that the material is exposed to that temperature, and a sufficient oxygen supply to insure destruction. Different materials require different kinds of incinerators.[18,20] For example, a rotary kiln incinerator can handle different kinds of solids, but can be expensive to operate and maintain. A fluidized bed incinerator is more limited in scope but can be effective if the waste is clearly defined and the bed is maintained. I was involved with liquid incinerators used in the U.S. to destroy our chemical weapons stockpile. Each of the incinerators was expensive and required careful monitoring of not only the facility itself but of the feed into it. I have been at meetings at which we focused entirely on technical issues related to safe functioning. Frankly, I believe that incineration is problematic unless at least one of the two following market conditions exist:

- For private investment, there needs to be a sufficient market to guarantee a large baseload and prices must be high enough to operate and maintain the equipment and training; and/or
- Public investment guarantees that the technology can be built, operated and maintained properly.

Currently, there is a market for land-based incineration that includes fixed facilities, and special mobile incineration facilities. In the U.S., using EPA reports from recent years, I estimate that about 7% of the hazardous waste generated is incinerated. However, pollution prevention has markedly reduced the market for incineration. Reporting in *Chemical Engineering News*, Kemsley[22] reviewed the construction of a new land-based incinerator by Clean Harbors, Inc. The $120 million facility is the first built in the United States in the twenty-first century! Kemsley attributes the lack of building to earlier overbuilding of incineration capacity. I do not disagree about the assertion that there was over-building. But I believe the reason the capacity has not been augmented is that companies decided that pollution prevention was a much better option than incineration and other disposal methods. What this new incinerator implies for me is that world demand is increasing to absorb idle capacity for some kinds of hazardous wastes. Also, the new facility demonstrates that expectations for emission reductions continue to increase. For example, the new El Dorado plant's emission requirements for mercury, cadmium, and lead are reductions of over 90% below existing plants, and 75% or more for dioxin and furan emissions.

For incinerating at sea, there is no current reason to assume that it will become a viable option, albeit one should never say never to an option that is plausible. The London Convention of 1972 has evolved and incineration at sea is considered ocean dumping because emissions fall onto ocean waters.[23] Hence, since the passage of the Ocean Dumping Ban Act of 1988 and the London Convention incineration of industrial and sewage sludge at sea is off the table.

International transfer of noxious production and waste management facilities

Globalization has made international movement more feasible. Media stories have focused on offshoring of presumably desirable activity. Many of the production processes and their wastes discussed here would be considered undesirable by many people and their governments. The long history of transfer of production was summarized and illustrated in Chapters 1 to 3. Here, I start in the 1980s when it became clear that concern about pollution was driving companies to a let-it-go policy. Among the various publications, Jeffrey Leonard's[24] *Pollution and the Struggle for the World Product* focuses on one of two key issues: Is there a comparative advantage for countries that have weak standards for toxins and waste management? In other words, is a weak regulatory regime replacing or supplanting cheap labor as a source of comparative advantage? Leonard uses Ireland, Mexico, Romania and Spain, concluding that location decision-making has become so complicated that he could not say that weak environmental requirements were an advantage. However, he noted that it was important in the case of some extremely dangerous facilities, in other words, noxious facilities. Furthermore, his case studies are no longer appropriate for evaluation in this second decade of the twenty-first century. Today, we would need to systematically evaluate Africa, South Asia, South America, and other mostly southern hemisphere nations. The work needs to be updated, but the hypothesis is relevant over a quarter of a century after Leonard's book was published.

The second key issue is international policy regarding movement of hazardous wastes across political boundaries. In 1989, a year after the publication of Leonard's book, the Basel Convention on the Control of Transboundary Movements of Hazardous Wastes and their Disposal was issued.[25,26] The Convention attempted to control the spread of hazardous wastes to developing nations in three ways:

- Companies should not export waste to a country that banned it;
- The exporting country should not allow exports if the import country does not consent to the import; and
- The exporting company should not export to a country that cannot properly manage the waste.

Spurred by blatant instances of what seemed to be unethical dumping practices, the Basel Convention was signed by over 20 countries and now has almost every nation in the world as a participant. Nothing, however, about this subject is simple. The United States was one of the first to sign and ratify it in 1990. But the U.S. has not implemented it. The U.S. is conspicuous as by the far the most populous and developed nation that has not legally implemented the Basel Convention. The idea has not been well received in some quarters in the U.S., primarily because it does not allow some existing practices. For example, it forbid some forms of recycling, that is, sending waste from one country to another as a raw material.[27] The State Department reports that the U.S. does not vote but sits in on meetings. I have strong opinions on lack of U.S. participation. For me it is a national embarrassment. Rather than using a lot of precious space for further argument, I strongly recommend Rebecca Kirby's 1994 law article as a good introduction to the U.S. political process that has stopped forward movement in the United States.[28]

A lot of anecdotal stories have been collected to suggest that a letting-it-go policy is harmful to human health and safety, as well as the environment. Some of these have been used to paint let-it-go as another form of colonialism. Beck[29] examined the possibility that cross-border movement increased industrial accidents, or alternatively that large corporations brought their progressive policies to other countries, and hence there would not be a major difference in health and environmental impacts. The data set was from industrial accidents during the period 1971–2000. The author found a disproportionate number of industrial accidents in poor countries and these were connected to worker illiteracy and dubious worker practices. This systematic study is rare. I periodically find other accounts, but few efforts at systematic studies.

Such studies are extremely difficult to conduct. For example, Vietnam[30–32] is a new hot economic zone with about 40% of its employment in manufacturing and a place where many international corporations are investing. With companies like Samsung, Nokia, and others manufacturing electronics in Vietnam, it is ironic that the U.S. has become its best customer for all the electronic products, and China is Vietnam's major supplier of materials and products. For Vietnam it means that the country has risen from one of the poorest countries in the world to a low middle income country. Yet, there have been incidents of toxic contamination in Vietnam.

Indonesia is a rival of Vietnam for development. Like Vietnam, it has had success stories, and yet there are incidents of concern in Vietnam. One of the most distressing is a blow-out in 2006 from a drilling operation in Sidoarjo, East Java, which has led to the emergence of a mud volcano that continues at a diminished rate in 2016. This mudflow has produced some of the worst pictures I have seen for many years. These two cases are anecdotal, and what we need is a country-by-country profile that systematically investigates the Leonard hypothesis and searches for incidents that were at the core of Beck's paper. The environmental–health–social and political implications of international migration of noxious activities is an open question at the country and local levels.

It follows from my previously expressed discomfort about focusing on anecdotal information and places that I have not personally experienced that I have chosen the lead industry as a case study because so much is known, and I have some personal experience with the case of lead. It represents, however, an extremely painful example of a let-it-go policy.

Case study: lead

Lead, like mercury (Hg), is a basic metal with an atomic number of 82. Like mercury, lead's uses go back thousands of years.[33] – glazes on prehistoric ceramics, kohl – a cosmetic used by the Egyptians, and the water pipes that some argue contributed to the decline of the Roman Empire. Fast forward to the 1950s, and I first became interested in lead because my parents bought a house that had been occupied by a lover of *National Geographic* magazine. He had cut out the maps and used them as wallpaper in what became my room. Immediately behind my pillow was a map of Australia, which was identified as a place that had a lot of lead. Being fascinated by maps and chemistry, I learned that lead was in our outdoor house paint, solder, gasoline, batteries, ammunition, cosmetics, type used by publishers and many other uses. When I consulted our year 1953 world atlas, I learned the U.S. had even more lead mining areas (big circle over Missouri), followed by Australia, and then Canada, Mexico, the USSR, and Burma (now Myanmar). For context, I also had a slab of asbestos in my room behind the radiator to protect against fire, and of course we had thermometers with mercury.

Only several decades later, did I understand the environmental hazards associated with mercury, lead, and asbestos, which at that time were in many consumer products. Lead was literally everywhere – air, water, soil, and inside. It can be carried many miles and settle down as dust or be carried by water from mines, refineries and manufacturing facilities. Also, as we all have learned it has been used in many products that we routinely use or had used, such as ceramics, pewter, paint, gasoline, and water from lead pipes.

Lead is a strong neuro toxin, affecting almost every organ, and it can cross the blood–brain barrier and access the central nervous system.[34,35] Children are at highest risk because they absorb more of it than adults; much of it goes to the brain, and their bodies do not have the same capacity to detoxify the metal and eliminate it in waste. As a result, in essence, there is no safe threshold for lead exposure. A recent study shows that U.S. adolescents from Chicago recruited during 1995–1997, matched to early childhood blood lead levels and followed to age 17 have significantly higher levels of impulsivity anxiety–depression, and body mass index (BMI), respectively, even after adjusting individual, household and neighborhood attributes.

One of the saddest afternoons I spent was when a neighbor told me that his son had been detected with high blood lead levels and had to be treated by chelation. He could not figure out where the exposure had occurred. After I put on my epidemiology hat, it took about an hour to figure out that the exposure had occurred when the outside of his house had been painted and the painter had left the windows wide open, allowing lead to fall inside the window sills and the surrounding floor. The young child must have crawled over and eaten quite a few lead paint chips. Unfortunately, this story has been repeated too many times.

To put the lead burden into context, I turn to a study by Lim et al. They conducted a comparative risk assessment of disease burden and injury attributable to 67 risk factors in 21 regions of the world.[35] The risk factors were grouped in 10 categories and the metric was estimated deaths and disability-adjusted years. At the top of the list of 67 factors were nutrition, smoking, high blood pressure, high BMI, alcohol, high blood pressure, tobacco smoking and secondhand smoke. Lead exposure ranked 17th of 67 with an estimated count of 674,000 in the year 2010. Furthermore, this is an increase from a 1990 estimate of 210,000. Even recognizing the imperfection of the data and the country-by-country estimates, this estimate is an intolerable amount of exposure.

Fast forward to the latter half of the twentieth century and the uses of lead have multiplied and spread from developed nations to every nation. The International Lead Association (ILA) monitors both the production and use of lead.[36] It characterizes itself as representing the world's large lead producing companies. It sees itself as working toward a sustainable future for the industry and with safe production, handling and use of lead. ILA's website presents stories about the successes of individual companies both economically and environmentally.

ILA reports show that in the year 2012 lead mine production was about 5.2 million tonnes a year up from 3.37 in 1990.[36] In 1990, Australia, the U.S., and China produced the most lead concentrate – 16.9%, 14.7%, and 9.3%, respectively. In 2012, the proportions of Australia and the U.S. declined to 12.5% and 6.7%, respectively, whereas China's proportion increased to 54.2%. World production increased 53%; Australia's climbed 14%, the U.S. production decreased, and China's increased almost nine times. ILA's database also provides data on lead refining, and other output measures. But not to lose the key point in the details, one important macro-scale piece of information is that big circle in my 1950s atlas that showed massive lead mining, refining and production in the U.S. and Australia, and little in China. That pattern has markedly changed (Figure 8.1). China produces more than U.S. and Australia combined. The only part of the lead lifecycle that is not dominated by China is recycling, which in fact is an indicator of the severity of the legacy described below.

Figure 8.1 Major lead production sites.

To bring this down from the macro to the micro scale, I am focusing on the United States, especially the southeast part of Missouri for the early history and then China along the South China Sea.

Lead and southeast Missouri

Missouri's official state mineral is galena, which says that lead is part of the history. For much of the late nineteenth and early twentieth century, Missouri was the global leader in lead production, that is, the big circle in my family atlas was primarily about Missouri. Actually, lead mining in Missouri began in the early eighteenth century with French explorers.

Lead was an important economic stimulator in Missouri, especially the southeast part of the state. For much of the last century, Missouri had the largest primary lead smelter in the U.S. at Herculaneum, and the largest secondary lead smelter in the world at Buick.

The Missouri Department of Natural Resources[37] has documented the role of lead mining in the state. The so-called "Old Lead Belt" and "New Lead Belt" are located in Washington County, Missouri, and adjacent areas, roughly 500 square miles. There also is a second approximately 2000 square mile "Tri-State" belt in southwest Missouri, and extending into nearby Kansas and Oklahoma. The third is the "Central" area around the Lake of the Ozarks (Figure 8.2).

Here, I focus on the Old Lead Belt, which included mostly St. Francois County in towns like Doe Run, Bonne Terre, Desloge, Park Hills, and Leadwood, as well as smaller operations in Washington and Madison Counties. During the period from 1864 through 1972, more than a dozen companies mined lead and zinc in this area, including St. Joe, ASARCO, St. Louis Smelting and Refining (later part of National Lead), and many others. St. Joe built a

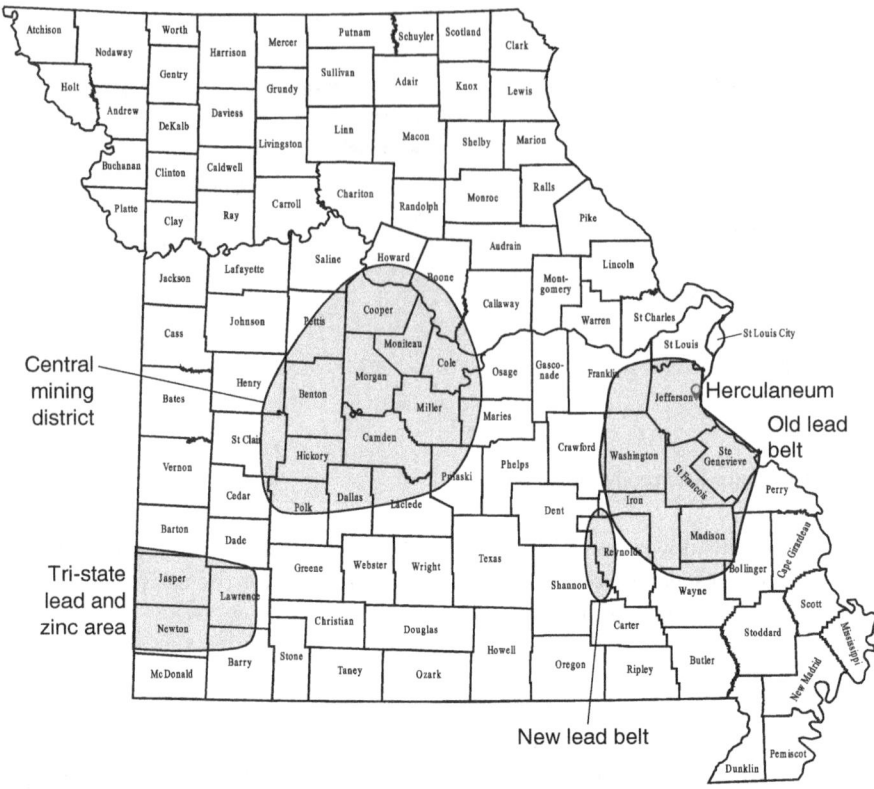

Figure 8.2 Missouri lead production areas.

lead smelter at Herculaneum in 1891, using galena as the primary ore. The towns were relatively prosperous.

In June 2016, MSNBC News[38] wrote a story on "life in Missouri's Fading Old Lead Belt." The story talks about the economic decline of the area and "environmental and that health hazards loom large" in the area that once was a major lead supplier. The story goes back to when French explorer Philip Francois Renault found lead in the area and mined it. It reports that only 3% to 5% of the material was lead, which has left a huge legacy of mine tailings in the area (I have seen sections of this). Lead mined from the area was used for ammunition and earlier versions of lead-acid batteries. The authors characterized the area as a "shadow of its former self," that is, poverty and a shortage of jobs, as well as health and environmental residues were rampant.

Piles of mining slag and waste represent an ongoing problem of wind-blown and waterborne lead. Efforts have been made to cover them with gravel, but the band aid solution does nothing for the legacy already in water bodies and on the land. This story and many others report cases of children that played on the slag piles and now have elevated blood lead levels. In other words, lead mining that started in the 1860s and ended in 1972 has left a legacy.

The Company Doe Run is a major producer of lead, copper and zinc concentrates. The company owned four mills, six mines, and a lead battery recycling plant in southeast Missouri, and facilities in the states of Washington and Arizona. It began in 1864 as the St. Joseph Lead Company, and built the lead smelter at Herculaneum in 1892 along the Mississippi River. That smelter operated until 2013 and was the last operating primary lead smelter in the United States.[39,40]

Its CEO Jerry Pyatt noted that in 2015 Doe Run spent $63 million on environmental protection and remediation projects, a good portion on remediation of the Herculaneum site. It also built a new water treatment plant at its Buick Mine and Mill, and invested in energy saving geothermal equipment. Also, natural resource damage claims have been made in the area that the company will need to address.[41]

One fascinating part of the ending of the story was the attack on the EPA by some writers for closing the smelter because the U.S., some argued, would need to rely on other countries to manufacture ammunition. (As noted below about 80% is used for lead-acid batteries and about 3% for ammunition.) A second fascinating factoid I was told was that China would entirely control the industry and from a public health perspective the U.S. was better off allowing Doe Run to produce lead from concentrated ore than China, which it was argued was not focused on health and safety. It is ironic that much of the lead mined in the Missouri lead belt went to lead-acid batteries that ended up in cars manufactured in Flint and in nearby Michigan. It seems that Flint and the Old Missouri lead belt share a common serious lead-related legacy.

Lead-acid battery and China

Among the various uses of lead, I focus on lead-acid batteries. The choice was easy. The International Lead and Zinc Study Group database for the period 2012 to early 2017 shows that batteries comprise 80% of the end use. This means that shot/ammunition, alloys, pigments, rolled and extruded products from lead, as well as cable sheathing, account for 20%.

The lead-acid battery was invented in the middle of the nineteenth century. It is inexpensive to produce, provides a strong boost to start motors, and has a high power-to-weight ratio. Many types of lead-acid batteries are used and progress has made them more effective and less dangerous. For example, a valve was added that along with other design changes has allowed the lead-acid battery to be adapted to be more flexible in regard to size and placement. This so-called "sealed" battery design causes hydrogen and oxygen produced in the battery to be recombined back to water, thereby reducing vibration, leakage, and size.

The lithium-ion is the main competitor, using two forms of lithium to obtain ion movement.[42,43] Lithium batteries are not risk free. Overheating and fire has resulted, which I suspect many readers of this chapter, including this author, have experienced in their computers and phones. Relatively high cost is a concern, but the price has been dropping as demand rises. Harrison believes that the growth of Tesla's electric cars and cell phones will lead to annual compound growth of 10.8% a year between 2015 and 2022.[42] However, for the immediate future the lead-acid battery is a key economic product.

China has become the largest producer, consumer as well as exporter of lead-acid batteries.[44] Measured by volt production, China's lead-acid battery industry grew over seven times during the period 1998–2011. The growth of the lead-acid battery industry in China is testimony to rapid urbanization and industrialization, and a reduction of poverty. The batteries are in cars, bicycles, communication, and energy storage devices. I have seen lists of hundreds of Chinese lead-acid battery producers. Almost 60% of the lead-acid batteries are produced in Jiangsu, Guandgong and Zhejiang provinces along the east coast, with Shanghai between them (Figure 8.3). Refining occurs further to west, for example, Henan is a major refining center located about 350 miles (559 km from Jiangsu), and Hunan another refining center is about 553 miles (890 km) from Zhejiang. Mining occurs further west in more rural locations across the country.

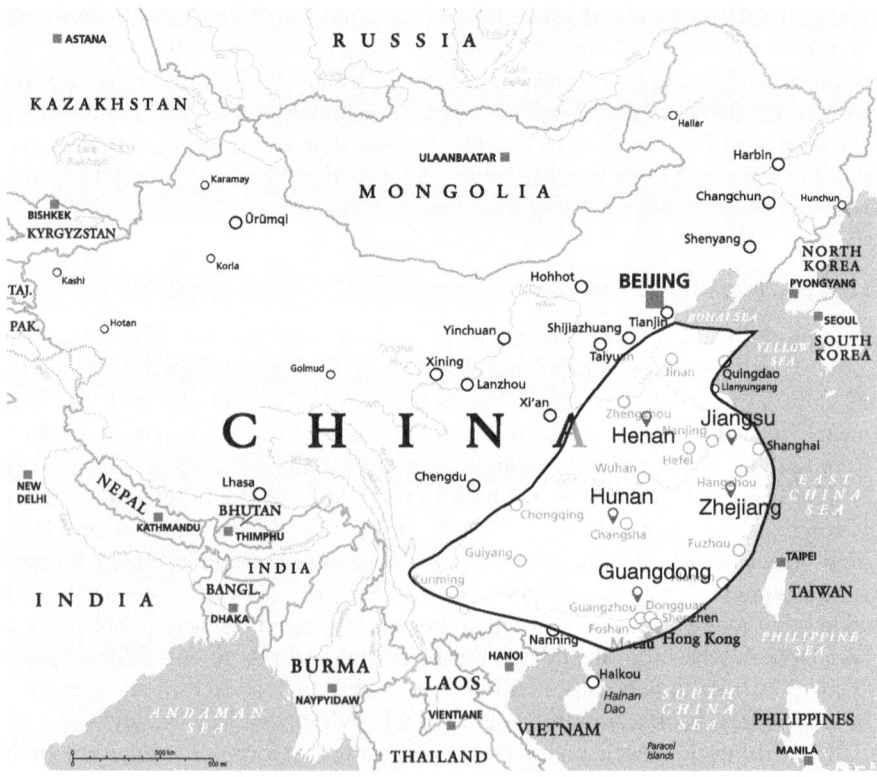

Figure 8.3 Coastal China lead production areas.

Lead-related pollution predictably has followed the jobs. Reports say that that only 30% of the lead-acid batteries are recycled, which means that 70% are not. Van der Kuijp et al.[45] report that many lead-acid batteries are discarded and recycling is by individuals with little training, including children working as part of their family business.

They report that from 2001 to 2007, 24% of Chinese children were poisoned by lead (lead level exceeding 100 ug/l is considered dangerous to children; toxicology data suggest a much lower threshold). This proportion compares to 16% the global average and less than 5% in the U.S. and EU. The Chinese numbers represent a decrease from higher proportions that existed until China banned leaded gasoline in 2000.

Reporting in the *New York Times* in 2011, LaFraniere[46] called lead poisoning in China the hidden scourge. She focuses the story on a small village in Zhejiang and traces the impact of a lead-acid manufacturing plant on local residents, especially large numbers of children as well as adults, many who are reported with blood lead levels that that are above a public health emergency level. The reporters note that some researchers indicate that one-third of Chinese children suffer from elevated blood lead levels. The story includes information about public displays of anger, government officials charged with not doing their jobs being arrested for failure to perform their duty, and other recriminations. With about 2000 factories and about 1000 battery recycling plants the ability of the government to control the contamination represents a daunting challenge. While I have concentrated on direct contact exposure for children, a great deal of the lead gets into water supplies and food.

In 2008, China's Ministry of Environmental Protection passed standards for the lead battery industry, and implemented them in 2009. In 2011, the Chinese government following inspections closed over 70% of the lead-acid factories in Zhejiang, Jiangsu, and overall suspended over 60% of the production in China. Ying Luo[47] called it a "clean up storm," with people not allowed to return to their homes if they lived closer than 500 meters from a factory. Not surprisingly, some of these people were ill from lead poisoning. Many lost their jobs in the lead plants. Companies that depend on battery use were scrambling for other options.

However, it is not as if the government can permanently shut down this industry and try to export the industry to other countries because China is the world's major manufacturer of electric vehicles, which have been rapidly expanding.[48] Writing in 2015, Daxue Consutling[49] asserts that demand for lead-acid batteries will grow by 16% through the year 2020.

Technological developments have improved lead-acid battery life, e-bikes will take over a good deal of the market, and presumably lithium batteries will begin to replace lead-acid ones. When I visited China in 2010, I spoke with a senior member of the Academy of Environmental Sciences who told me that trading low cost for lead pollution is no longer acceptable, and he used the lead-acid battery to illustrate the need for all countries to deal with the lead-acid battery problem. Daxue Consulting[49] suggests that in the near term future

China's goal is to replace the lead-acid battery with other methods of powering automobiles and motor bikes, and they will use the lead-acid battery more for storing energy as part of the electrical grid, where the batteries can be monitored, recycled and reused.

Analysts[50–52] expect growth of the lead-acid battery industry in China to stop because of the government's reaction to lead poisoning and the slow entry of the lithium battery into the marketplace. They believe smaller, inefficient, and polluting facilities will be closed and larger, more efficient, and safer plants will emerge as well as a modern recycling industry.

Where does it end?

In a span of less than two decades, China became the largest lead-acid battery producer in the world and received a devastating reminder of the public health implications of lead. I do not believe that the world community will continue to tolerate the public health impacts for the indefinite future. China has tightened the rules, closed many smaller facilities and centralized all parts of the industry, especially the manufacturing. China has been clear about not wanting to be the world's producer of lead-acid batteries. Plants are opening in Vietnam, Malaysia, and Bangladesh, which will bring production closer to Australian lead sources. But the health costs will continue in these places, unless this industry changes its operations, as is being done in the United States and more recently China.

Eventually, lead will be treated more like asbestos and mercury, that is, limited mining, controlled recycling rather than discarding, and production in controlled and managed places, and then only under the most regulated procedures. The public health carnage associated with lead needs to stop and it is hard for me to imagine that it will not. Letting it go elsewhere can no longer be consider ethically acceptable, if it ever was.

International examples

Let-it-go policies of international companies include buying out or building noxious facilities, as well as illegal waste dumping scandals. In 2013, Smithfield Foods, the world's largest producer of pork, was acquired by Chinese food company Shuanghui International.[53,54] It was the largest Chinese buyout of an American firm – for $7.1 billion – estimated as 30% above the real market value of Smithfield Foods. Also, it was reported that the company would own 25% of the pigs raised in the United States after the buyout, which will be exported to China.[54,55] This merger has several aspects from dietary habits of ever-growing Chinese middle-class to national food security.[55,56] Environmental problems caused by factory farming and Concentrated Animal Feeding Operations (CAFOs) constitute an important part of the deal as well. Smithfield Foods is also known for massive manure production and being the major polluter among agribusinesses for toxic releases into water supplies.[54,57] It is argued that Chinese will enjoy the final products, while Americans will be the ones who have to deal with environmental problems in the long run.[53,55]

In 2016, Taiwan's Formosa Plastics Group, said it will invest $9.4 billion to build petrochemical plants in Louisiana.[58,59] The company expects faster approval under the administration of President Donald Trump and to start operation one year ahead of schedule. The Chairman of the company said in an interview to Reuters, "Trump has said his priority is the U.S. economy first, and then environmental protection" and "at least the obstacles will be fewer" regarding project approval.[60] Interestingly, the Taiwanese government refused to require an environmental impact assessment for a domestic investment.[61] The Trump administration was criticized for a tolerant attitude for permits and regulations in environmentally sensitive industries.[60,62]

Illegal shipment of hazardous wastes from industrialized European countries to underdeveloped West African countries has been an ongoing issue since the 1980s.[63] In 1988, illegal shipments of toxic waste of two Italian companies to Koko, Nigeria, led to a major scandal. It was revealed that an agreement was made with the owner of the land to store 8000 drums of toxic waste for $100 a month.[64,65] Initially, the Italian government insisted that the waste was harmless. Only after the potential danger and toxic level in the waste were proven, did the Italian government agree to remove it.[66] A similar incident happened in 2006, when an international company named Trafigura, dumped its toxic waste in Abidjan, Côte d'Ivoire. The reason for dumping was said to be that the company refused to pay for proper disposal in the Netherlands.[67] Later on, it was reported that the ship contained two tonnes of hazardous hydrogen sulfide.[68] The authorities announced that 17 people died and about 100,000 people got sick because of the incident.[67,69] In 2010, a Dutch court fined the company, the captain of the ship, and an employee.[69,70] Ten years after the incident, it was reported that the toxic cleanup is not finished and compensation had not been completely paid.[71]

Final thoughts

Letting it go means different things to different populations. In the United States, in recent years it has meant many jobs have been lost leading to public anger and finger-pointing and promises to reshore some of this industry. In regard to noxious industry, in many cases it has been pressure from the public and not-for-profit groups that have made it economically, socially, and politically unacceptable to keep the industry anywhere near people, or it was just another reason to eliminate an unprofitable part of its portfolio.

Companies and government officials are loathe to admit that public anger is as important as the geological, hydrological, and other physical attributes of a possible site. But having been cursed, yelled at and otherwise verbally abused at hearings in the United States, I understand why finding a site elsewhere is a good option, or so it seems at first. The problem is that even if the let-it-go option is legal, desirable for a company's bottom line, and a solution for a government that does not know what to do in charged political environments, the long-term reality is that there is a class of noxious facilities that go beyond the Basel Convention and banning dumping at sea that need internationally agreed upon rules of engagement. Even if a company disassociates itself from a business so it can turn back to its core mission and explore more profitable ones, these companies and governments that support them need guidelines and sometimes rules that guide and restrict relocation for the sake of future generations.

References

1 Weintraub J (producer), Anderson P (director) *Soldier*. United States. Warner Brothers, 1998.
2 Coopersmith J. Nuclear Waste in Space? August 22, 2005. https://ntrs.nasa.gov/archive/nasa/casi.ntrs.nasa.gov/19780015628.pdf. Accessed March 27, 2018.
3 Quora. Why Don't We Dispose of Nuclear Waste Into Outer Space? www.quora.com/Why-dont-we-dispose-of-nuclear-waste-into-outer-space. Accessed March 27, 2018.
4 Burns R, Causey W, Galloway W, Nelson R. Nuclear Waste Disposal in Space. NASA Technical Paper 1225. Marshall Space Flight Center, Alabama, May 1978.
5 World Nuclear Association. Radioactive Wastes – Myths and Realities. February 2016. www.world-nuclear.org/information-library/nuclear-fuel-cycle/nuclear-wastes/radioactive-wastes-myths-and-realities.aspx. Accessed March 27, 2018.
6 Golan M, Globus Y. (producers), Furie S. (director) *Superman IV*. United States. Warner Brothers, 1987.
7 Lapp R. One Answer to the Atomic-Energy Puzzle – Put the Atomic Power Plants on the Ocean. *New York Times Magazine*. June 4, 1972, 20–21–23, 80, 84, 90.
8 Dames & Moore. Siting Considerations for Offshore Nuclear power Plants. Los Angeles, CA, Dames & Moore, September 1973.
9 Novak M. The American Plan to Build Nuclear Power Plants in the Ocean. www.smithsonianmag.com/history/the-American-plan-to-build-nuclear-power-plants-in-the-ocean-27801262/. Accessed March 27, 2018.
10 Stover D. Floating Nuclear Power Plants: China is Far from the First. *Bulletin of the Atomic Scientists*. https://thebulletin.org/floating-nuclear-power-plants-china-far-first9522. Accessed March 27, 2018.
11 Prosser M. China's Curious Dream of Floating Nuclear Plants on the Ocean. February 17, 2016. https://singularityhub.com/2016/02/17/chinas-curious-dream-of-floating-nuclear-plants-on-the-ocean/#sm.0000eqikfzmz3e8ot5o23wmidysz3. Accessed March 27, 2018.
12 *Economist*. Nuclear Power. All at Sea. April 26, 2014. www.economist.com/news/science-and-technology/21601231-researchers-find-advantages-floating-nuclear-power-stations-all-sea. Accessed March 27, 2018.
13 MIT. Offshore Small Modular Reactor (OSMR): An Innovative Plant Design for Societally Acceptable and Economically Attractive Nuclear Economy in a Post-Fukushima, Post-9/11 World. DOI:10–1115/SMR2014–3306. Web.mit.edu/news/office. Accessed May 12, 2017.
14 Stauffer N. A New Look for Nuclear Power. June 24, 2015. http://news.mit.edu/2015/new-look-floating-nuclear-power-0624. Accessed May 12, 2017.
15 Greenberg M. *Explaining Risk Analysis: Protecting Health and the Environment*. New York, Earthscan/Routledge, 2016.
16 Costner P, Thornton J. *Playing with Fire: Hazardous Waste Incineration*. Washington, DC, Greenpeace, 1990.

17 Greenpeace. New Zealand. Why Ban Incineration. December 5, 2006. www.green peace.org/new-zealand/en/campaigns/toxics/incineration/why-ban-incineration/. Accessed March 27, 2018.

18 U.S. EPA. Combustion. https://archive.epa.gov/epawaste/hazard/tsd/td/web/html/ combustion.html. Accessed May 10, 2017.

19 Bond D. At-Sea Incineration of Hazardous Wastes: The Risk Is Yet To Be Justified. *Environmental Science & Technology.* 18, 5, 148a-152a, 1984.

20 Science Advisory Board, U.S. EPA. *Report on the Incineration of Liquid Hazardous Wastes by the Environmental Effects, Transport and Fate Committee.* A 101, Washington, DC. April 1985.

21 Sandstedt J. Lessons form At-Sea Incineration, Inc. paper presented at conference – the Practical Politics of Hazardous Waste Management, Transcending the Siting Dilemma, Washington, DC, May 6, 1987.

22 Kemsley J. New Hazardous Waste Incinerator Comes Online. *Chemical & Engineering News.* 95, 14, 34–35, April 2017. https://cen.acs.org/articles/95/i14/New-hazardous-waste-incinerator-comes.html. Accessed March 27, 2018.

23 U.S. EPA. Learn About Ocean Dumping. www.epa.gov/ocean-dumping/learn-about-ocean-dumping. Accessed March 27, 2018.

24 Leonard J. *Pollution and the Struggle for the World Product.* Melbourne, Cambridge University Press, 1988.

25 Basel Convention. Parties to the Basel Convention on the Control of Transboundary Movements of Hazardous Wastes and Their Disposal. www.basel.int/Countries/Status ofRatifications/PartiesSignatories/tabid/4499/Default.apsc. Access May 15, 2017.

26 U.S. Department of State. Basel Convention on Hazardous Wastes. https:// 2001–2009.state.gov/g/oes/envr/e18124.htm. Accessed May 15, 2017.

27 Hanson D. International Hazardous Wastes Treaty Worries U.S. Industry. *Chemical & Engineering News.* February 26, 1996. 22–24.

28 Kirby R. The Basel Convention and the Need for United States Implementation. *Georgia Journal of International and Comparative Law,* 24, 281–305, 2014.

29 Beck M. The Risk Implications of Globalisation: An Exploratory Analysis of 105 Major Industrial Incidents (1971–2010). *International Journal of Environmental Research and Public Health.* 13, 3, 309–330.

30 Phung Mai DT, Barbour-Lacy E. An Introduction to Vietnam's Import & Export Industries. www.vietnam-briefing.com/news/introduction-vietnams-export-import-industries.html/. Accessed March 27, 2018.

31 World Bank. World Development Indicators. Data.worldbank.org. Accessed May 15, 2017.

32 Toan P, Nguyen M, Nguyen H. Energy Supply, Demand and Policy in Vietnam with Future Projects. *Energy Policy.* 39, 11, 6814–6826, 2011.

33 U.S. EPA. Learn About Lead. www.epa.gov/lead/learn-about-lead. Accessed March 27, 2018.

34 Winter A. The Consequences of Early Childhood Lead Exposure for Adolescent Health in a Prospective Birth Cohort. *American Journal of Public Health.* 107, 9 1496–1501, 2017.

35 Lim et al. 2012. A Comparative Risk Assessment of Burden of Disease and Injury Attributable to 67 Risk Factors and Risk Factor Clusters in 21 Regions, 1990–2010: A Systematic Analysis for the Global Burden of Disease Study, 2010. *Lancet.* 380, 9859, 2224–2260.

36 ILA. International Lead Association. Responsibility. www.ila-lead.org/responsibility. Accessed March 27, 2018.

37 Department of Natural Resources, State of Missouri, Missouri Lead Mining History by County. http://dnr.mo.gov/env/hwp/sfund/lead/mo-history-more.htm. Accessed May 4, 2017.

38 Hoste B, Hoogwaerts R. Life in Missouri's Fading Old Lead Belt. www.msnbc.com/msnbc/life-missouris-fading-old-lead-belt. Accessed May 4, 2017.

39 Jones T. Last U.S. Lead Smelter to Shutter in Ore-Rich Missouri. www.bloomberg.com/news/articles/2013-12-23/last-u-s-lead-smelter-to-shutter-in-ore-rich-missouri. Accessed March 27, 2018.

40 Worstall T. The Last Lead Smelter in the U.S. Closes Because the Hippies Won. *Forbes*. December 23, 2013. www.forbes.com/sites/timworstall/2013/12/23/the-last-lead-smelter-in-the-us-closes-because-the-hippies-won/#4b804aea642b. Accessed March 27, 2018.

41 Doe Run. Message from the CEO. http://sustainability2016.doerun.com/message-from-the-ceo/. Accessed March 27, 2018.

42 Harrison K. 5 Factors Powering the Lithium-Ion Technology Boom. www.inc.com/kate-l-harrison/5-factors-powering-the-lithium-ion-battery-boom.html. Accessed March 27, 2018.

43 Hunt T. Is There Enough Lithium to Maintain the Growth of the Lithium-Ion Battery Market? June 2, 2015. www.greentechmedia.com/articles/read/is-there-enough-lithium-to-maintain-the-growth-of-the-lithium-ion-battery-m#gs.g1FpuA4. Accessed March 27, 2018.

44 International Lead Association. Lead Production & Statistics. www.ila-lead.org/lead-facts/lead-production--statistics. Accessed March 27, 2018.

45 Van der Kuijp TJ, Huang L, Cherry C. Health Hazards of China's Lead-Acid Battery Industry: A Review of its Markets Drivers, Production Processes, and Health Impacts. *Environmental Health*. 2013. https://ehjournal.biomedcentral.com/articles/10.1186/1476–069X-12–61. Accessed May 16, 2017.

46 LaFraniere S. Lead Poisoning in China: The Hidden Scourge. *New York Times*. www.nytimes.com/2011/06/15/world/asia/15lead.html. Accessed May 16, 2017.

47 Luo Y. Production of Lead-Acid Batteries Indefinitely Suspended in China. June 22, 2011. www.bloomberg.com/news/articles/2011-11-15/china-shuts-90-of-lead-acid-battery-makers-association-says. Accessed March 27, 2018.

48 International Lead and Zinc Study Group. End Uses of Lead. www.ilzsg.org/static/enduses.aspx?from=1. Accessed March 27, 2018.

49 Daxue Consulting. Lead Acid Batteries in China Still Growing. http://daxueconsulting.com/lead-acid-batteries-china-still-growing/. Accessed March 27, 2018.

50 Woods L. Global and China Lead-Acid Battery Industry Report 2016–2018 – Johnson Controls is the World's Largest Lead-Acid Battery Producer with 15.7% Market Share. September 12, 2016. https://globenewswire.com/news-release/2016/09/12/871232/0/en/Global-and-China-Lead-acid-Battery-Industry-Report-2016-2018-Johnson-Controls-is-the-World-s-Largest-Lead-Acid-Battery-Producer-with-a-15-7-Market-Share.html. Accessed March 27, 2018.

51 Tian X, Wu Y, Gong Y, Zuo T. The Lead-Acid Battery Industry in China: Outlook for Production and Recycling. *Waste Management & Research*. 33, 11, 986–994, 2015.

52 Mancini L. China Does Not Want to Become the Lead-Acid Battery Manufacturer for the Whole World. www.linkedin.com/pulse/china-does-want-become-lead-acid-battery-manufacturer-lorenzo-mancini. Accessed March 27, 2018.

53 Hauter W. Food Industry Deals Hurt Consumers and the Environment. *New York Times*. 2013. www.nytimes.com/roomfordebate/2013/06/02/smithfield-china-and-the-calculus-of-transnational-deals/food-industry-deals-hurt-consumers-and-the-environment. Accessed August 4, 2017.

54 Mercola J. United States Is Turning Into a Factory Farm for China, With Devastating Environmental Consequences. *Mercola*. 2016. http://articles.mercola.com/sites/articles/archive/2016/12/06/cafos-ruin-farm-land-waterways.aspx. Accessed August 4, 2017.

55 Halverson N. How China Purchased a Prime Cut of America's Pork Industry. *Reveal*. 2015. www.revealnews.org/article/how-china-purchased-a-prime-cut-of-americas-pork-industry/. Accessed August 4, 2017.

56 Campbell C. China May Not Have Enough Arable Land to Feed Its People. But Big Changes Are Coming. *Time.* 2016. http://time.com/4455462/china-agriculture-food-security/. Accessed August 4, 2017.

57 Environment North Carolina Research & Policy Center. Corporate Agribusiness and the Fouling of America's Waterways: The Role of Large Agribusiness Companies in Polluting our Rivers, Lakes and Coastal Waters. 2016.

58 Ting-Fang C. Investing $9.4bn in U.S. Ethane Facility. *Nikkei Asian Review.* 2016. https://asia.nikkei.com/Business/AC/Investing-9.4bn-in-US-ethane-facility. Accessed August 4, 2017.

59 Lin A, Cho S, Koh A. Taiwan's Formosa Seeks U.S. Permit for $9.4 Billion Investment. *Bloomberg.* 2017. www.bloomberg.com/news/articles/2017-02-20/taiwan-s-formosa-seeks-u-s-permit-for-9-4-billion-investment. Accessed March 27, 2018.

60 Hung F. Formosa Expects Faster OK for U.S. Petrochemical Plant Under New EPA Chief. *Reuters.* 2017. www.reuters.com/article/us-fpcc-investment-idUSKBN1683SI?il=0. Accessed August 8, 2017.

61 Chou C. Formosa Plastics to Boost U.S. Investment. *The China Post.* 2017. www.chinapost.com.tw/business/company-focus/2017/02/18/491860/formosa-plastics.htm. Accessed August 4, 2017.

62 Kanason E. Trump Fuels a Petrochemical Boom in America. *OilVoice.* 2017. https://oilvoice.com/Opinion/2795/Trump-Fuels-a-Petrochemical-Boom-in-America. Accessed August 8, 2017.

63 Brooke J. Waste Dumpers Turning to West Africa. *New York Times.* 1988. www.nytimes.com/1988/07/17/world/waste-dumpers-turning-to-west-africa.html?pagewanted=all. Accessed August 4, 2017.

64 Buck S. In the 1980s, Italy Paid a Nigerian Town $100 a Month to Store Toxic Waste—and It's Happening Again. *Timeline.* 2017. https://timeline.com/koko-nigeria-italy-toxic-waste-159a6487b5aa. Accessed August 4, 2017.

65 Greenpeace. The toxic ships: The Italian hub, the Mediterranean and Africa. 2010. Rome, Italy. www.greenpeace.org/italy/Global/italy/report/2010/inquinamento/Report-The-toxic-ship.pdf. Accessed March 27, 2018.

66 Jenkins L. After dumping on Nigeria, Italy takes it all back. *Washington Post.* 1988. www.washingtonpost.com/archive/politics/1988/09/04/after-dumping-on-nigeria-italy-takes-it-all-back/. Accessed August 4, 2017.

67 Greenpeace. The Toxic Truth: About a Company called Trafigura, a Ship Called the Probo Koala, and the Dumping of Toxic Waste in Côte d'Ivoire. 2012. The Netherlands. https://reliefweb.int/report/c%C3%B4te-divoire/toxic-truth-about-company-called-trafigura-ship-called-probo-koala-and-dumping. Accessed March 27, 2018.

68 Leigh D, Hirsch A. Papers Prove Trafigura Ship Dumped Toxic Waste in Ivory Coast. *Guardian.* 2009. www.theguardian.com/environment/2009/may/13/trafigura-ivory-coast-documents-toxic-waste. Accessed August 4, 2017.

69 CNN. Dutch Court Issues Fine Over Ivory Coast Waste Dump. *CNN.* 2010. www.cnn.com/2010/WORLD/europe/07/23/netherlands.ivory.coast.trafigura/index.html. Accessed August 4, 2017.

70 BBC News. Trafigura found guilty of exporting toxic waste. *BBC News.* 2010. www.bbc.com/news/world-africa-10735255. Accessed August 4, 2017.

71 Mail Online. 10 years on, I. Coast Toxic Waste Clean-up 'Not Complete'. *Mail Online.* 2016. www.dailymail.co.uk/wires/afp/article-3749395/10-years-I-Coast-toxic-waste-clean-not-complete.html. Accessed August 4, 2017.

Part III

Tools and coping with siting noxious facilities in the early twenty-first century

9 Tools

Introduction

The two objectives of this chapter are:

- Introduce a set of tools that have been suggested and sometimes used as decision aids in siting noxious facilities; and
- Offer selected references to keep up with information about noxious facility siting.

With regard to the first objective, some tools are discussed in Parts I and II, but there are many more process and analytical tools than have been mentioned in the first eight chapters. All 24 types of tools presented here are in the academic and business literature, and most have been highly touted. Some have made a difference in siting cases and others have not lived up to their hype. I divided them into three broad categories: social/political, economic/business, and environmental/health. Some fit into multiple categories, for example, mapping fits into all three categories.

The 24 tool types are presented in alphabetical order within each of the three categories. Following the name of the tool, the presentation provides information about:

1   Objective of the tool;
2   Typical applications;
3   Prerequisites for its use;
4   Assessment of its use to site noxious facilities; and
5   Citations for follow-up (an average of three per topic).

Two caveats are in order before you consult these brief reviews. First, my original list of tools included more than 75. Rather than present an encyclopedia of tools, I was selective and combined some tools that are related, for example, scenarios and charrettes.

Second, these tools are risk-informing aids, not decision-making ones. Here are several illustrations from my own work that illustrate their limitations. For example, the first one in the list below is the analytical hierarchy process (AHP). A widely touted and applied tool, I last used it at the request of a company that was considering relocating its international headquarters. We organized several groups of participants who developed a hierarchy of key siting criteria, and then we applied these to more than a dozen cities in Europe and Asia. The net result was that London was the highest rated site, even after we changed the weights on the criteria and conducted sensitivity analysis on the outcomes. The organization was happy, we were pleased with the way the tool worked in practice, and the organization started to use the results.

On June 23, 2016, the United Kingdom and Gibraltar voted by a slim majority to leave the EU. London ranked number 1 in our analysis: it was part of the EU, was politically stable, was accessible from so many locations, a place with many communications options, and had other advantages. If I went back and recalculated the number with the UK not in the EU and what that implies, I doubt it would even be among the top three choices. In short, we had a good question to answer, the tool was a good match to the objective, and the study went along smoothly, the client liked the results, that is, until an event that we had not considered as plausible changed the assumptions used in the analytical hierarchy process model.

Another example is the application of optimization models, which have had limited success when applied to noxious facility siting. For example, I wrote a book that presents numerous illustrations of the use of optimization models, including optimizing the use of labor and capital, while protecting human health and the environment. The models have worked in regard to more efficient use of labor and equipment in hospitals, transportation assignments, and many others.[1] Yet, when I applied optimization to siting landfills and incinerators, I learned some lessons about the limitations of this class of models. The State of New Jersey's policy was that each of the 21 counties should manage its own household waste. This policy made no sense. We developed a simple optimization model that sought to minimize waste management costs subject to the need to manage all the waste. We tried a variety of sites of varying sizes and both incinerators and landfills.

The major finding was that it was possible to cluster new sites and use existing sites rather than require every county to have a site. The cost savings were clear, and fewer new sites would be needed. Indeed, government and business chose sites near those we used in the model. On the other hand, we were severely criticized by some local governments that said we chose sites that were not zoned for waste management. Technically, they were correct. Few, if any, local governments zone any of their land for waste management and those that do usually allocate small parcels of land that are isolated from roads and rail lines. Hence, we chose areas zoned for industrial purposes with good transportation connections. In fact, in our opinion, many of these large parcels were zoned industrial to avoid unwanted residential land uses. We opened a political boil about affordable housing in this waste management siting process and faced some angry elected officials. They wrote letters to us and to our senior administration criticizing us for even conducting the study and using taxpayer money to conduct it.

Nevertheless, in hindsight this tool worked to show that the state policy needed rethinking. In fact, Governor James Florio cancelled some of the proposed county-specific sites, which was a successful outcome as far as we were concerned. Overall, this was largely a success story, but the key message is that the closer the results came to specific locations, the more they were challenged. My book is now out of date. I suggest *Fuzzy Stochastic Optimization*[2] for an up-to-date book on the use of this class of tools. Yet, as shown by this example, optimization modeling involving sites and constraint mapping (Chapter 7) is intended to eliminate sites and suggest a few sites for further on-site research. But they do not eliminate every site and elected officials in towns that that have not been eliminated typically will do whatever they can to attack the results as long as they are on the possible site list. With these examples as context, the chapter turns to summaries of the 24 tools.

Social/political tools

Analytical hierarchy process

1   A multistage process that develops decision criteria, gathers data about them, and uses input by interest groups to rate all the options, and then compares them.
2   Widely used when a multiplicity of data types, attributes, and interest groups are involved.
3   Project leaders should have experience with the AHP process and be able to facilitate exchanges about different criteria, weights, and metrics.
4   This tool is not as widely used in siting as I would expect, and yet it can be successful in siting cases where the parties are trying to juggle different kinds of data and can use the flexible AHP process to understand the implications of changing assumptions.
5   For follow-up I recommend the Wasil, Golden[3] special issue of case studies, and two classics by Thomas Saaty, the developer of the tool.[4,5]

Collaborative problem solving and other alternative dispute resolution

1   A group of eight to 12 representative stakeholders work with a professional mediator/advisor and staff to improve understanding of information, options, and uncertainty with the goal of providing suggestions to the decision-makers or actually make decisions on behalf of the decision-makers.
2   Applied when there is time and resources to form a representative group to understand information and either offer some suggestions, a consensus decision, or even a unanimous decision.

3   Group leader must be unbiased, be able to recruit and retain staff, and most important keep the stakeholders on path to a decision.
4   High stakes commitment to a group and process that could lead to a win-win for all the key parties or to a failure.
5   Good case-grounded literature is available about consensus-building, including Gray[6]; Hartnett[7]; Saint, Lawson[8]; and Susskind, McKearnen, Thomas-Larmer, which is my favorite.[9]

Demographic segmentation analysis, including environmental justice analysis

1   Using published census, marketing, and survey data at neighborhood (census block and tract), municipal, county, state and regional scales to classify the resident population into groups that are germane to decisions under consideration. This is particularly important if the population has been classified as part as environmental justice population under Executive Order 12898 (see Chapter 4).
2   Widely applied by business and government, however, often lacking in sophistication in regard to data quality, representativeness of the data, and statistical analysis, leading to overly simplistic and potentially misleading results. Some analysis is required by environmental impact assessments and some state assessments.
3   Already prepared data can be easily downloaded with only limited preparation. More demanding analyses require gathering, weighting, and then analyzing raw data using EJ screen, factor, cluster and other statistical tools.
4   Increasingly used in siting decisions, especially for environmental justice cases. There is tendency for parties to make exaggerated claims, especially at public meetings, often causing disruption and consternation. It is critical that the parties have access to professionally analyzed and unbiased results, rather than information sorted to be ammunition in value-laden debates. Any concern about EJ-related impacts must be addressed.
5   The literature is broad, and readers need to deal with survey, analysis and data display methods that if not properly treated can undermine validity and understanding of the results. See Greenberg and Weiner's survey paper[10] and Dillman, Smyth, and Christian.[11] In regard to statistical analysis users need analysts who can use factor and cluster analysis. I recommend Afifi and Clark's text.[12] In regard to environmental justice I suggest the U.S. Environmental Protection Agency's Environmental Justice Screening and Mapping Tool, which is a wonderful teaching tool.[13]

Mapping

1   The objective is to communicate economic, environmental, health, infra-structure, political, social and other information to groups with different backgrounds, which is likely to require overlaying maps to show places that

are most likely to benefit from economic growth and places that may bear the brunt of the risk. One key is to not rush through the presentation and a second is to listen to responses.

2   Use has markedly expanded with the availability of high speed mapping tools and will increase as big data efforts multiply.

3   A great deal of experience is required to produce maps that inform rather than mislead participants. Serious problems can result from misuse of colors, scales, labels and other symbols on maps. Participants may feel that they have been deceived by poor maps and/or may misunderstand the data because of the way it has been displayed.

4   Mapping is an opportunity to build trust by preparing draft maps and then discussing what is and is not shown, the scale, the colors and other map elements. This approach should be a requisite in noxious facility siting as a way of increasing knowledge and building trust.

5   A great deal has been written about how to and not display data: see, for example, Kraak and Ormeling[14] and the Hazards and Vulnerability Research Institute[15] to see how data are created and displayed. See also EJ screen under demographic segmentation analysis & environmental justice (above).

Political analysis

1   A process to trace and understand the history of political pressure to change policy and practice.

2   Typically done after a problem is apparent, rather than before, which is unfortunate because sometimes a problem can be prevented or at least reduced if it has been anticipated.

3   Historians, political scientists, and attorneys typically conduct the research.

4   Developers have begun to learn that they need to do political research before a problem erupts that eliminates some or all of their siting options and their credibility. Failure to engage in this type of research has been a missed opportunity in facility siting cases.

5   Some excellent examples exist, especially around the civil rights movement. McAdam, Tarrow, and Tilly[16] and Morris.[17] The National Advisory Commission on Civil Rights[18] published a report in 1967 that I consider to be the single best post hoc example of an impactful political analysis.

Risk communications: listening and speaking

1   Focus groups, media messages, public meetings, surveys, websites, and other ways that allow parties to provide information to each other and to obtain information from each other.

2   Widely used to provide information to stakeholders, less often used to listen to concerns, which is a missed opportunity to get at what really concerns people.

3   Multiple tools, each of which requires appropriate expertise to use.
4   Risk communications is now a centerpiece of every siting case. Sometimes the parties listen and talk, other times they spend too much time trying to persuade stakeholders and insufficient time listening to other views. The literature suggests that risk communication can increase competence-related trust but has little impact on values and that trying to convince people with different values and priorities to put aside these aside can back-fire, leading to stronger opposition.
5   I would start with Cho, Reimer, and McComas's[19] edited book. Then I would review Lundgren and McMakin.[20] For a classic on the need for and process of risk communication, see Sandman.[21]

Scenarios and charrettes

1   Translation of plausible ideas into text and then into maps, plans and other ways of expressing ideas in computers and wooden or cardboard models in order to stimulate reactions and creative input.
2   Only reasonable to use when designers can offer multiple options that can be altered.
3   Designers must have the ability to express their ideas in multiple formats that persons with limited experience can understand and manipulate;
4   Eminently good idea in siting cases when alternatives can be expressed as two- and three-dimensional computer maps, physical blocks and other forms to help stakeholders get a realistic idea of what impacts may occur, and even potentially more effective if there is room for them to adjust the designs and learn to trust each other as result of the exercise.
5   See Saint, Lawson,[22] Linstone, Turoff,[23] Lindsey, Todd et al.,[24] and U.S. EPA[25] for a variety of uses of scenarios and charrettes.

Social network analysis

1   Qualitative and quantitative methods that trace the formation, expansion, and shrinkage of groups that take positions on important policy issues.
2   Historically, this was done by attending meetings, talking with people and reading written reports. Today, the web is the source of much of the data and new computer-based tools exist to map a network and determine who is at the center of the network, which is important for all parties.
3   Simple networks can be mapped without advanced training or tools, complex ones require professionally trained experts with a computer package and skill.
4   Could be valuable to organizations proposing a project as a way of deter-mining who to invite to participate in meetings and negotiations about a noxious facility site.
5   An important new set of tools added to pre-existing network analysis tools. See De Nooy, Mrvar, Batagelj[26]; Mislove, Marcon, Gummadi, et al.[27]; and Cohen-Cole and Fletcher[28] for some tool discussions and cases.

Economic/business tools

Benefit and cost analysis (BCA or CBA)

1 A set of economic-driven methods used to calculate and then compare the economic benefits and costs of a set of projects/programs in the short run and throughout the life cycle of the activity;

2 Widely used to support capital investment and complex infrastructure programs and many other public and private programs, including siting.

3 Strong background in economic theory and previous use of the method is essential.

4 A decision-aid tool in siting, even if tools are not applied as suggested in the literature. Most decision-makers have some training in BCA and some have personal experience and are comfortable with the approach. Proponents frequently encounter criticism on ethical grounds that the tool ignores the value of life and other attributes that are hard to monetize. Methods like cost-effectiveness analysis have been introduced to emphasize other objectives (see also ecosystem evaluation and resource damage assessment, green accounting).

5 Readers interested in this topic should keep up with issues of the *Journal of Benefit-Cost Analysis*,[29] which is the flagship journal of the Society for Benefit-Cost Analysis, and see the U.S. Environmental Protection Agency[30]; and Zerbe and Bellas's[31] book.

Compensation

1 Payments aimed at increasing support for an action. In regard to siting, compensation may be used if negotiations occur or when a site is taken by the federal or state government without negotiations. Compensation may be in the form of taxes, payments in lieu of taxes, grants, purchases, loans, property value guarantees, infrastructure additions, and other payments.

2 Routine in transactions, but more iffy when introducing new sites or enlarging existing ones via clamping.

3 Parties must be able to bring a list of ideas to the table and be able to estimate the cost of these options, and understand other parties' ideas.

4 Use of compensation becomes much more challenging when a proposed site is noxious. The pool of sites is likely to be quite limited, and consequently compensation typically is paired with volunteerism, that is, if there are volunteers then compensation can be designed as an auction process in which the local government asks for monetary compensation for agreeing to host a facility, or the party can skip the auction and begin bargaining with multiple parties. In cases, where the local and state government cannot disapprove the site, negotiations can be useful to demonstrate that the body(ies) imposing the site do respect the local interest.

5   Highly touted in the literature and some success, but much opposition in noxious facility cases. I suggest two classics on the subject: O'Hare, Bacow and Sanderson[32] for a thoughtful presentation of compensation in the context of local opposition, and Himmelberger, Ratick, and White[33] for an early empirical study of what factors predict success and failure.

Economic efficiency models

1   Beginning with simple optimization models, researchers have offered mixed integer and linear and non-linear models that minimize costs. Then tools have added multiple objectives, uncertainty, lotteries, volunteers, and temporal dimensions as part of multbjective designs.
2   The approach has been widely touted as a way of focusing on economic efficiency and incorporating environmental health and other constraints into decision-aid tools.
3   Strength and experience with quantitative economic modeling is required, including linear, mixed integer and linear, dynamic, stochastic, dynamic, and other models.
4   Many applications suggested in the literature, but payoffs measured by facilities sited have been limited. The real world poses many ethical, cultural, political and other issues that are hard to incorporate into even the most sophisticated multbjective models. Environmental justice issues represent a major challenge for this set of approaches. Tools seem to work more effectively when siting decision options are limited and tool is combined with GIS.
5   I suggest the following for those wanting to read more about where thinking about these models has been moving: Farahani, Steadiesefie, and Asgari[34]; Shin and Sure,[35] and Santore.[36]

Ecosystem evaluation and natural resources damage

1   Typically, value estimates are based on field studies, mapping, expert judgment and modeling.
2   Requires expertise in ecology, biology, economics, mapping, and spatial simulation modeling.
3   Placing a "price tag" on natural resources and the value of damaging them is, in essence, imputing human values to biological, chemical and physical processes. Estimating the overall value of a forest as a carbon recycling system, a surface and underground water system, and many others represent challenging exercises in obtaining accurate data, and assigning explicit and implicit values of cumulative public benefits. For example, the economic values of a scenic view are hard to measure, except in places where there commercial recreational facilities that maintain records. A natural resource damage assessment is normally more confined spatially and temporally so that experts can focus on a replacement value of a

specific groundwater supply to the public, and the value of fish that have been decimated because of contamination, or the value of a scenic view. Ecological risk analyses is an excellent framework to avoid making bad siting decisions that could backfire in major natural resource damage costs in the future.

5   For follow-up I suggest the following: U.S. EPA[37]; U.S. EPA Ecosystem Services[38]; Jacobs, Dendoncker, and Keune[39]; and Environmental Law Institute.[40] Also see the websites of NOAA, U.S. Department of Interior, as well as U.S. EPA for presentations about natural resources damage.

Green accounting and green budgets

1   Using accounting to monitor environmental impact costs associated with business decisions and to include estimated costs in decision-making for strategic planning.

2   A direct challenge to business to rethink how development impacts the environment and human health and safety. Case studies show that the idea has caused business to consider the issue. However, many businesses and government organizations resist the idea, especially formal accounting changes called for by green accounting. A few empirical studies show that investors view the practice as positive evidence that an organization is concerned about exposure to human health and safety costs and regulations, and that good environmental performance is associated with good economic performance.

3   Requires that accounting systems and experts monitor inputs and outputs that otherwise many never be considered, and accept the idea of rethinking what is and is not accounted for. Green accounting, in essence, is a step toward incorporating sustainable practices into doing business.

4   Siting represents an ideal use of green accounting. There is no logical reason why private and public organizations could not monetize the traditional economic factors and the more recent set of issues as part of up-front, back-end and contingent costs, albeit the estimates are imperfect. Unofficially, the leaders of some major corporations are doing this as testified to by the principles that they have adopted. However, some may not be explicitly monetizing all of the factors that could be considered.

5   For more information I suggest the following: U.S. EPA[41]; Al-Tuwaijri, Christensen, Hughes[42]; and examine the websites of some organizations that have embraced the concept: AT&T, Mazda, Toshiba, and the U.S. Department of Energy.

Regional economic impact analysis

1   Quantitative tools used to estimate economic impacts of proposed programs and projects at regional, state, and local scales, and for periods ranging from a few months out to 20 years.

2   Typically applied to large complex projects that are likely to have major impacts over a decade or more.
3   Strength in quantitative economic modeling required, specifically background with econometrics, input–output analysis, and if possible Regional Economic Modelling, Inc. as well as computable general equilibrium models.
4   Most economic estimates are made with simplistic tools that miss indirect economic impacts. This leads to underestimates of impacts, and can also lead to exaggeration of impacts by site proponents. Some recent studies have begun to use current models. Parties appreciate the information but are likely to challenge it when they go against their values. Cost and expertise are a challenge.
5   For an overview of the interface of examining the economic impact of hazard events see Greenberg, Lahr, Mantell[43]; Miller, Blair[44]; Partridge, Rickman[45]; Rose, Liao[46]; and Treyz.[47]

Environmental/health tools

Chemical alternatives assessment

1   A process used to compare human health, environmental impacts, and sustainability impacts of different chemicals. Includes green chemistry, sustainable designs, and other processes that require explicit analysis of alternatives.
2   Focuses on avoiding bad chemical substitutions, including such considerations as types of hazards, life cycle exposures, technical demands of substitution, economic and other trade-offs among options.
3   Expertise in chemistry, exposure assessment, toxicology, epidemiology, risk assessment are essential to compare options.
4   A core part of pollution prevention by private and public organizations discussed in Chapter 5. A key outcome should be making choices that do not make a substitution that leads to worse or equally bad outcomes, of which we have seen many such as Methyl tert-butyl ether (MTBE) as a gasoline additive, which did raise the octane number of gasoline but caused water pollution.
5   I suggest the National Research Council's review[48]; a report by the California Department of Toxic Substances Control[49]; and Lavoie, Heine, Holder et al.[50]

Ecological and environmental footprint analysis

1   A system of accounting originally designed to measure and compare resource use by country. More recently, the idea has spread to organizations, which allows them to compare their performance with counterparts. Footprint analysis is part of pollution prevention and sustainability.

2   Users begin with the general idea of comparing water, developed land, food energy, materials use, waste production and transportation resource use and then the impact of these on biologically productive land, water bodies, forests needed to absorb carbon emissions, buildings, roads and other built lands, and on overall biodiversity.

3   An old idea that has been rapidly developing with the creation of metrics, and scientists who can move the ideas from global ideas about consumption down to scales where government and business can act to reduce their footprint.

4   An opportunity for government and private organizations to learn about themselves and manage their organizations in line with the idea of shrinking their footprint. Such analyses could be a powerful part of information provided as part of a siting plan.

5   A rapidly developing literature. I would start with Nicholson, Chamber, and Green[51;] European Commission[52]; and Moore[53]; and for remediation projects related to siting and reuse, I would read U.S. EPA.[54]

Environmental Impact Assessment (EIS)

1   A sequential set of analyses and interactions required by federal, as well as by some state and local governments in order to assess the environmental consequences of projects/programs. Economic, environmental justice, and social justice are also included in an EIS but not as thoroughly done as environmental assessments.

2   Required by the National Environmental Policy Act of 1970 of all major federally supported projects and those that require federal permits, licenses, and other legal approvals.

3   Multidisciplinary teams of experts in physical and social sciences required to gather, assess and integrate data-based findings and advise decision-makers about a preferred option, a no action alternative and other options.

4   Required for all U.S. government projects and major projects that require federal government permits, financial assistance other major federal government support. Some states have their own forms of EIS to require noxious facilities to prepare an assessment, and Newark, New Jersey was the first municipal government in the United States to require a report to determine if there is a potential environmental justice issue. For noxious facilities reporting varies from thousands of pages to a checklist.

5   The most direct reference is to search environmental impact statement under the U.S. Environmental Protection Agency. For projects related to the Department of Energy, Defense, and every other agency, please go directly to the department/agency website to see their processes. Note, however, that the EPA will in the end be required to comment on everyone. See also the journal, the *Environmental Impact Assessment Review*; and see Greenberg[55] for a historical perspective.

Exposure assessment

1   A hybrid of environmental science and epidemiology used to recreate and predict the magnitude, duration and other metrics of hazardous exposures to humans, especially workers.
2   Exposure assessment estimates the number of exposed people and workers, primarily in the workplace environment, but also in homes and in some outdoor environments.
3   Background in epidemiology, biology, chemistry, physics, environmental health and toxicology are essential in order to model and measure exposures.
4   Exposure assessment has been a major addition to the toolkit available for risk assessors to realistically forecast the human health and safety implications of siting and non-siting designs focusing on sources of hazards and pathways of exposure from source to potentially exposed people.
5   I recommend several sources, beginning with the U.S. Occupational Safety and Health Administration (OSHA)[56] and EPA.[57] For a case study, I suggest case the World Health Organization's (WHO) mercury exposure study.[58]

Green chemistry and green engineering

1   Use of chemistry and chemical engineering to reduce exposure to hazards throughout the life cycle of the product.
2   Green chemistry and engineering are at the core of efforts to prevent pollution and to constantly reduce emissions and exposure by many progressive companies and nations. The basic focus is on designing products, manufacturing processes, picking transportation alternatives, and marketing programs that reduce human exposure, and increase sustainability.
3   Combinations of creative science and engineering working together have made enormous changes that have reduced hazards and increased economic benefits throughout the lifecycle of a product.
4   Success in green chemistry and engineering reduce the need for noxious sites, their footprint, public concern, waste that must be managed, and data show increased credibility of the organizations, including increasing their economic value.
5   The places to look are the American Chemical Society (ACS)[59]; and U.S. EPA.[60,61] I would also read Vallero and Brasier.[62]

Hazard mitigation analysis

1   A U.S. federally mandated planning program requiring states, multi-county regions and many local governments to determine what hazards exist, how they can be managed, and what to do to reduce the likelihood of an accident.

2 Required by FEMA as a part of the message that local government must assume the responsibility through planning for reducing major hazard events, especially chronic damage events.

3 Teams of experts with experience in hazards, risk, mitigation and resilience are critical to guide local, regional and state planning.

4 Every hazard mitigation plan that I have read includes some information about noxious facilities. With big data capability increasing, it will become increasingly possible to link facility plans with mitigation plans.

5 Some helpful documents are available to guide mitigation planning, such as Hazards and Vulnerability Research Institute[63]; Federal Emergency Management Agency (FEMA)[64]; Godschalk, Beatley, Berke et al.[65]

Health impact assessment

1 An ordered process to assess human health and safety impacts and offer solutions.

2 A tool that fits into a niche of small- to medium-sized projects that fit between the environmental impact statement and back of the envelope processes.

3 Formal training in the HIA process is essential, as well as subject matter expertise.

4 HIA is not required but could be applied in some siting cases, most likely clamping cases (Chapter 6). The advantage is that the process heavily depends on participation of different stakeholders and forces the participants to clearly state what are their public health concerns and how to address them without relying on full scale risk assessment models, although models could be incorporated into the process.

5 Some excellent introductions and reference sites have recently appeared, such as Kemm[66]; Birley[67]; SOPHIA[68]; International Association for Impact Assessment,[69] the leading global network on impact assessment www.iaia. org/; and NACCHO.[70]

Life cycle assessment (LCA)

1 Procedures to estimate the environmental consequences and more recently social ones by measuring inputs and outputs of energy and other materials.

2 Widely used during the early twenty-first century to learn about resource use through the stages of mining, transportation, production, and other steps of producing, managing, selling, distributing, and post-sales use and management.

3 Widespread expertise essential to understand, map out and trace resource use. Closely related to green engineering, accounting and chemistry.

4 Noxious industries have learned a great deal about where they use resources in their processes. As a result of regulations, LCA, and related analyses, organizations have introduced pollution prevention, closed sites and let

some products or parts of products move elsewhere, and challenged organizations to consider a variety of siting and non-sting options.

5   I would go directly to the websites of the U.S. Environmental Protection Agency[71]; the Global Development Research Center[72]; and the International Organization for Standardization (ISO 14040 and 14044).[73] I suggest reading the Levi-Strauss[74] life cycle analysis of jeans, which is an eye opener for those that do the laundry.

Resilience systems analysis (RSA)

1   A fascinating opportunity to break out from traditional risk management that has been limited to selected consequences and limited solutions. RSA pushes out into hazards like earthquakes, world-wide financial shocks and others that can devastate multiple social, environmental and political systems. Teams of non-experts and experts explore these shocks and how to manage and recover from them.

2   Begins with current or recent risk landscapes and anticipates how these might evolve given local, regional, national and world-wide events. RSA takes some of these risks and examines them in detail through risk analysis. In the process, the group follows how the original risk event (e.g., earthquake, financial crisis, energy shortages) will cascade through economic, social, political system impacting people, business, infrastructure, services and so many other attributes.

3   Without doubt, participants must be able to understand different kinds of events and their impacts, but they must be representative of and in tune with different cultures, patterns of behavior and the area's history.

4   An opportunity and challenge for those proposing siting and non-siting options. A resilience systems analysis may persuade local populations and officials that public and private organizations are able to cope with hazards and thereby increase credulity. On the other hand, the opposite could result and low probability and low consequence events could become a major concern because they were studied. An important goal for a siting project is to develop a list of and explore seasonal, periodic, and idiosyncratic shocks and their management. (See also hazard mitigation analysis.) Like sustainability and life cycle analysis, resilience systems analyses can easily become a buzzword that attracts different and opposing opinions and leads to words rather than to actions.

5   A good place to start is the OECD's[75] guidelines for resilience systems analysis, and Folke's[76] article.

Sustainability impact assessment (SIA)

1   A self-declared "soft" policy tool directed at requiring organizations to explicitly consider social, economic and environmental consequences of their actions. This process is notably broader and more demanding of ability

to compare than are cost and benefit analysis, the environmental impact assessment, environmental justice, regulatory and other impact assessment tools.

2   Developed and promoted by the Organization of European Development (OECD) to push organizations into as broad and future-oriented perspectives as possible.

3   Demands analysts with broad training and open minds to be able to consider a plethora of considerations in decision-making.

4   Demands that decision-makers consider a breadth of consequences resulting from siting and non-siting options. Given the broad set of criteria, developers must be able to find non-monetary methods to compare apples and oranges. Toward this end, the analytical hierarchy process is suggested (see above), as it features the ability to compare different kinds of metrics.

5   The best reference is the Organization for Economic Cooperation and Development (OECD), which consists of 35 mostly European and North American nations,[77] and I suggest two EPA-sponsored National Research Council reports about sustainability.[78,79]

Scanning

Staying in the moment is a good way of living, but scanning for trends reduces the chances of missing something important that is coming and needs to be considered. After repeatedly making mistakes that could have been avoided, the famous comedian Oliver Hardy coined the catch-phrase – "Well, here's another nice mess you have gotten me into." Ollie was rebuking his sidekick, Stan Laurel, after Hardy suffered various insults, including turning into a chimpanzee ("Dirty Work"), and being forced into a duel by a jealous husband ("The Fixer Uppers"), among others. Hardy's messes were laughable, but he seemed to repeatedly put himself at risk because he could not manage to anticipate.

Chapter 3 shows what happens when organizations picked sites that started as lemonade and ended as bitter lemons. Could they have avoided picking those sites or picked less bad ones? Given past practice, maybe I expected too much of them. But today post-failure rationalization is harder to justify, especially when failure comes with loss of credibility, economic, social and political penalties.

Chapters 2 and especially 3 in this book exemplify creating Laurel and Hardy-like messes. Chapter 2 summarizes the theory, and Chapter 3 the painful evidence. The following five references provide the background for understanding what analysts believed and acted upon 50 to 100 years ago, and unfortunately still act upon in some places:

Pred A.[80] *The Spatial Dynamics of U.S. Urban Industrial Growth, 1800–1914. Interpretive and Theoretical Essays.*
Smith D.[81] *Industrial Location.*
Weber A.[82] *Theory of the Location of Industries.*

Greenhut M.[83] *Plant Location in Theory and Practice.*
Tarr J.[84] *Devastation and Renewal: An Environmental History of Pittsburgh and Its Region.*

Looking forward is a challenge because of the complexity that confronts industrial planners. Chapter 10 addresses the issue of coping with increasing complexity. Here, I have listed the 12 websites I consult most often and the five I routinely consult to see what is happening that might lead to new site requirements, or to non-siting options, including clamping. Those that I check quite a bit are marked with an asterisk. My most productive search expression is "industrial expansion."

- *New York Times
- *Wall Street Journal
- *Bloomberg Business News
- *Forbes
- *Reuters
- Washington Post
- Fox News
- CNBC News
- Al Jazeera Media
- International Trade Administration
- The Economist
- Chemical and Engineering News

These are all economic-oriented citations. In regard to social, human health and environmental changes, I routinely consult the following websites, and use "industrial pollution" and "industrial facility siting" as keys:

- U.S. Environmental Protection Agency
- American Chemical Society
- American Petroleum Institute
- U.S. Department of Energy
- U.S. Nuclear Regulatory Commission

I also have set up a Google Scholar search to look for articles under "industrial location," "industrial siting," "noxious facility siting" and "location." These searches tend to yield papers about optimum siting and all too often sites that do not fit the definition of noxious I use. Nevertheless, if you open the site and then sort by date, you will keep track of recent publications.

Final thoughts

The science of finding sites is steadily advancing. There are more data and more tools supported by fast computers to circulate and examine data, simulate impacts, and obtain public input. I am persuaded that the vast majority of organizations already understand the value of the green tools and budgeting processes and will drag their less inclined colleagues along with them. This means that the fundamental tools challenge is to use all the information and tools more effectively than in the past. This is a Mount Everest-equivalent challenge.

For me, the major challenge in noxious facility siting is building and maintaining trust. As I was writing this paragraph, I received a call from a journalist who asked a few questions about a paper I had written about nuclear power plant location. With over 40 years' experience responding to questions from reporters, I could quickly tell that her goal was to determine if I could be trusted. Was that study credible or not was the unasked question. After asking about the survey methods, and then the analysis, she asked me if I had ever received funding from the DOE, DOD, DHS, EPA, states, companies, and foundations. The answer to all her questions was yes. I was funded by the so-called bad guys and good guys, which made her more uncomfortable. She could not fit me into a convenient category of source. Could the study be trusted and could my remarks be trusted. I did not appear in her story, so I guess that I flunked the trust test.

It is in this part of facility siting – building trust – where we need to do a lot better than in the past, even with all our tools and high speed computers. The data and the mathematical models can get us to the point where we can eliminate some uneconomical and unsafe sites. This brings us to where the real debate starts, which is over values, preferences and quality of life right now and in the future. I have my favorite tools from among the list of 24, and they are not the ones that are based on deep or even medium level science. They are what many people call the soft science ones, for example, negotiations using charrettes and scenarios, going to meetings and not only listening but being empathetic, and most important using collaborative problem solving with a tool like the analytical hierarchy process or a much simpler consensus-building process.

References

1 Greenberg M. *Applied Linear Programming for the Socioeconomic and Environmental Sciences.* New York, Academic Press, 1978.
2 Wang S, Watada J. *Fuzzy Stochastic Optimization: Theory, Models and Applications.* New York, Springer 2012.
3 Wasil E, Golden B. Public-Sector Applications of the Analytical Hierarchy Process. *Socio-Economic Planning Sciences.* Special Issue. 252, 87–165, 1991.
4 Saaty T. *Mathematical Principles of Decision-making.* Pittsburgh, RWS Publications, 2009.
5 Saaty T. *The Analytical Hierarchy Process.* New York, McGraw Hill, 1980.
6 Gray B. *Collaborating: Finding Common Ground for Multiparty Problems.* San Francisco, CA, Jossey Bass Publisher, 1989.
7 Hartnett T. *Consensus-Oriented Decision-Making: The CODM Model for Facilitating Groups to Widespread Agreement.* Gabriola Island, BC, Canada, New Society Publishers, 2011.
8 Saint S, Lawson J. *Rules for Reaching Consensus: A Modern Approach to Decision Making.* Amsterdam, Wiley, 1994.

9 Susskind LE, McKearnen S, Thomas-Larmer J. *The Consensus Building Handbook: A Comprehensive Guide to Reaching Agreement.* SAGE Publications, 1999.

10 Greenberg M, Weiner M. Keeping Surveys Valid, Reliable, and Useful: A Tutorial. *Risk Analysis.* 34, 8, 1362–1375, 2014.

11 Dillman DA, Smyth JD, Christian LM. *Internet, Mail, and Mixed-Mode Surveys: The Tailored Design Method.* 3rd ed. New York, Wiley, 2008.

12 Afifi A, Clark V. *Practical Multivariate Analysis.* Boca Raton, FL, Taylor & Francis, 2012.

13 U.S. Environmental Protection Agency. EJSCREEN: Environmental Justice Screening and Mapping Tool. www.epa.gov/ejscreen.

14 Kraak M-J, Ormeling F. *Cartography: Visualization of Geospatial Data.* 3rd ed. New York, Routledge, 2013.

15 Hazards and Vulnerability Research Institute. Social Vulnerability Index for the United States 2006–10. http://webra.cas.sc.edu/hvri/products/sovi.aspx.

16 McAdam D, Tarrow S, Tilly C. *Dynamics of Contention.* New York, Cambridge University Press, 2001.

17 Morris AD. *The Origins of the Civil Rights Movement: Black Communities Organizing for Change.* New York, Free Press, 1984.

18 National Advisory Commission on Civil Disorders. *Report of the National Advisory Commission on Civil Rights,* 1967. www.eisenhowerfoundation.org/docs/kerner.pdf.

19 Cho H, Reimer T, McComas K, eds. *The SAGE Handbook of Risk Communication.* Thousand Oaks, CA Sage, 2014.

20 Lundgren R, McMakin A. *Risk Communication: A Handbook for Communicating Environmental Safety, and Health Risks.* 5th ed., 2013. Hoboken, NJ, Wiley-IEEE Press, 2013.

21 Sandman P. *Responding to Community Outrage: Strategies for Effective Risk Communication.* 1993. www.psandman.com/book.htm.

22 Saint S, Lawson J. *Rules for Reaching Consensus: A Modern Approach to Decision Making.* Amsterdam; San Diego: Pfeiffer & Co., 1994.

23 Linstone H, Turoff M. *The Delphi Method Techniques and Applications.* http://is.njit.edu/pubs/delphibook/delphibook.pdf. Published 2002.

24 Lindsey G, Todd J, Hayter S, Ellis P. *A Handbook for Planning and Conducting Charrettes for High Performance Projects.* 2009. www.nrel.gov/docs/fy09osti/44051.pdf.

25 U.S. EPA. Example Exposure Scenarios Assessment Tool. https://cfpub.epa.gov/ncea/risk/recordisplay.cfm?deid=85843.

26 De Nooy W, Mrvar A, Batagelj V. *Exploratory Social Network Analysis with Pajek.* 2nd ed. New York, Cambridge University Press, 2011.

27 Mislove A, Marcon M, Gummadi K, Druschel P, Bhatacharjee B. Measurement and Analysis of Online Social Networks. In: *IMC'07, October 24–26.* San Diego, CA, 2007.

28 Cohen-Cole E, Fletcher JM. Is Obesity Contagious? Social Networks vs. Environmental Factors in the Obesity Epidemic. *J Health Economics.* 27, 5, 1382–1387, 2008.

29 Society for Benefit-Cost Analysis, *Journal of Benefit-Cost Analysis.*

30 U.S. Environmental Protection Agency. Overview of Economic Analysis at the EPA. www.epa.gov/environmental-economics/overview-economic-analysis-epa. Accessed March 28, 2018.

31 Zerbe R, Bellas A. *A Primer for Benefit-Cost Analysis.* Northampton, MA, Edgar Elgar Publishers, 2006.

32 O'Hare M, Bacow L, Sanderson D. *Facility Siting and Public Opposition,* New York, Van Nostrand Reinhold, 1983.

33 Himmelberger J, Ratick S, White A. Compensation for Risks: Host Community Benefits in Siting Locally Unwanted Facilities. *Environmental Management.* 15, 5, 647–659, 1991.

34 Farahani R, Steadiesefie M, Asgari N. Review Article: Multiple Criteria Facility Location Problems: A Survey. *Applied Mathematical Modeling.* 24, 1689–1709, 2010.

35 Shin G, Sure H. A Review of Hierarchical Facility Location Models. *Computers & Operations Research*. 34, 2310–2331, 2007.

36 Santore R. Noxious Facilities, Environmental Damages, and Efficient Randomized Siting. *Environmental Resource Economics*, 57, 101–116, 2014.

37 U.S. EPA. Natural Resource Damages: Assessments. www.epa.gov/superfund/natural-resource-damages-assessments. Accessed March 28, 2018.

38 U.S. EPA Ecosystem Services. www.epa.gov/eco-research/ecosystem-services. Accessed March 28, 2018.

39 Jacobs S, Dendoncker N, Keune H. *Ecosystem Services: Global Issues, Local Practices*. New York, Elsevier, 2013.

40 Environmental Law Institute. Natural Resource Damage Assessment. Gulf of Mexico Restoration and Recovery. www.restorethegulf.gov/natural-resource-damage-assessment-nrda. Accessed March 28, 2018.

41 U.S. EPA. An Introduction of Environmental Accounting as a Business Management Tool: Key Concepts and Terms, EPA 742-R-95-001. https://archive.epa.gov/p2/archive/web/pdf/busmgt.pdf. Accessed March 27, 2018.

42 Al-Tuwaijri SA, Christensen T, Hughes KE. The Relations Among Environmental Disclosure, Environmental Performance: A Simultaneous Equations Approach. *Accounting, Organizations and Society*. 29, 447–471, 2004.

43 Greenberg M, Lahr M, Mantell N. Understanding the Economic Costs and Benefits of Catastrophes and Their Aftermath: A Review and Suggestions for the U.S. Federal Government. *Risk Analysis*. 27, 1, 83–96, 2007.

44 Miller R, Blair P. *Input-Output Analysis: Foundations and Extensions*. New York, NY, Cambridge University Press, 2009.

45 Partridge M, Rickman D. Regional computable general equilibrium modeling: a survey and critical appraisal. *International Regional Science Review*. 21, 3, 205–248, 1998.

46 Rose A, Liao S-Y. Modeling Regional Economic Resilience to Disasters: A Computable General Equilibrium Analysis of Water Service Disruptions. *Journal of Regional Science*. 45, 1, 75–112, 2005.

47 Treyz G. *Regional Economic Modeling: A Systematic Approach to Economic Forecasting and Policy Analysis*. Boston, MA, Kluwer Academic Publishers, 1993.

48 National Research Council. *A Framework to Guide Selection of Chemical Alternatives*. Washington, DC, National Academy Press, 2014.

49 California Department of Toxic Substances Control. Alternatives Analysis. 2010. www.dtsc.ca.gov/SCP/AlternativesAnalysisGuidance.cfm. Accessed March 28, 2018.

50 Lavoie E, Heine L, Holder H, Rossi M, Lee R, Connor E, Vrabel M, DiFiore D, Davies C. Chemical Alternatives Assessment. Enabling Substitution to Safer Chemicals. *Environmental Science & Technology*. 44, 9244–9249, 2010.

51 Nicholson R, Chamber N, Green P. Ecological Footprint Analysis as a Project Assessment Tool. *Engineering Sustainability 156*. Issue ES3, 139–145, 2003.

52 The European Commission. Analysis of Existing Environmental Footprint Methodologies for Products and Organizations: Recommendations, Rationale, and Alignment. 2011. www.ec.europa.eu/environment/eussd/pdf/Deliverable.pdf.

53 Moore D. Global Footprint Analysis. Report about the San Francisco-Oakland-Fremont, CA Metropolitan Statistical Area. June 30, 2011. www.footprintnetwork.org/content/images/uploads/SF_Ecological_Footprint_Analysis.pdf. Accessed March 28, 2018.

54 U.S. EPA. Methodology & Spreadsheets for Environmental Footprint Analysis (SEFA). https://clu-in.org/greenremediation/methodology/.

55 Greenberg, M. *The Environmental Impact Statement After Two Generations: Managing Environmental Power*. New York, Routledge, 2011.

56 U.S. Occupational Safety and Health Administration (OSHA) www.osha.gov/SLTC/etools/respiratory/changeschedule_exposure.html.

57 U.S. EPA's exposure assessment guidance. https://we.epa.gov/risk/guidelines/exposure-assessment.

58 World Health Organization's (WHO) www.who.int/foodsafety/publications/chem/mercuryexposure.pdf.

59 American Chemical Society (ACS) https://search.acs.org/search?q=green+chemistry&client=acs_r2&output=xml_no_dtd&proxystylesheet=acs_r2&sort=date%3AD%3AL%3Ad1&entqr=3&oe=UTF-8&ie=UTF-8&ud=1&site=acs&filter=p&x=0&y=0. Accessed March 28, 2018.

60 U.S. EPA. www.epa.gov/greenchemistry. Accessed March 28, 2018.

61 U.S. EPA. www.epa.gov/green-engineering. Accessed March 28, 2018.

62 Vallero D, Braser C. *Sustainable Design: the Science of Sustainability and Green Engineering.* Hoboken, NJ, John Wiley & Sons, Inc, 2008.

63 Hazards and Vulnerability Research Institute. Social Vulnerability Index for the United States 2006–10. http://webra.cas.sc.edu/hvri/products/sovi.aspx.

64 Federal Emergency Management Agency (FEMA). *Multi-Hazard Mitigation Planning Guidance Under the Disaster Mitigation Act of 2000.* Washington, DC, FEMA, 2008.

65 Godschalk D, Beatley T, Berke P, Brower D, Kaiser EJ. *Natural Hazard Mitigation: Recasting Disaster Policy and Planning.* Washington, DC: Island Press, 1999.

66 Kemm J. *Health Impact Assessment: Past Achievement, Current Understanding, and Future Progress.* New York, Oxford University Press, 2013.

67 Birley M. *Health Impact Assessment: Principles and Practice.* London, Earthscan, 2011.

68 The Society of Practitioners of Health Impact. SOPHIA Assessment. 2014. http://hiasociety.org/.

69 International Association for Impact Assessment. Health Impact Assessment Blog. http://healthimpactassessment.blogspot.com/.

70 NACCHO. Planning for Healthy Places with Health Impact Assessments. http://advance.captus.com/planning/hia2/home.aspx.

71 U.S. EPA. (EPA.gov).

72 Global Development Research Center (gdrc.org).

73 International Organization for Standardization (ISO 14040 and 14044).

74 Levi-Strauss. Life Cycle Assessment of Jeans. 2007. www.levistrauss.com/sustainability/planet/lifecycle-assessment/. Accessed March 28, 2018.

75 OECD. Guidelines for Resilience Systems Analysis: How to Analyze Risk and Build a Roadmap to Resilience. 2014. www.oecd.org/doc/ResilienceSystemsAnalysis.FINAL.pdf.

76 Folke C. Resilience: The Emergence of a Perspective for Social-Ecological Systems Analysis. *Global Environmental Change.* 16, 257–267, 2006.

77 Organization for Economic Cooperation and Development (OECD). Sustainability Impact Assessment: An Introduction. 2016. www.oecd.org/greengrowth/48305527.pdf.

78 National Research Council. *Sustainability and the EPA.* Washington, DC, National Academy Press, 2011. www.nap.edu.

79 Kavanaugh M, Abbott S, Allen D, Amar P, Brooks B, Burke I, Crittenden J, Fava J, Gilman P, Greenberg M, Hutson A, Kling C, Matthews H, Petrovskis E, Suh H, Taylor A, Yosie T. Committee on Scientific Tools and Approaches for Sustainability. *Sustainability Concepts in Decision-Making.* Washington, DC, National Academy Press, 2014.

80 Pred A. *The Spatial Dynamics of U.S. Urban Industrial Growth, 1800–1914. Interpretive and Theoretical Essays.* Cambridge, MA, MIT Press, 1966.

81 Smith D. *Industrial Location.* New York, John Wiley and Sons, 1971.

82, Weber A. *Theory of the Location of Industries.* Chicago, University of Chicago Press (translation of German edition written in 1909), translated from German by Friedrich C.

83 Greenhut M. *Plant Location in Theory and Practice.* Chapel Hill, University of North Carolina Press, 1956.

84 Tarr J. *Devastation and Renewal: An Environmental History of Pittsburgh and Its Region.* Pittsburgh, PA, University of Pittsburgh Press, 2003.

10 Coping with siting and non-siting options

Introduction

Chapter 10 has three objectives:

- Summarize four key lessons learned from earlier chapters;
- Suggest how traditional economic and risk analysis approaches are necessary but not sufficient when the issue is siting noxious facilities;
- Illustrate the challenge of coping with siting in today's rapidly paced environment with two illustrations: the chronic problem of managing garbage in large cities; and a surprising policy change that could increase the pool of available noxious facility sites.

Lessons learned

During the last century, we learned a great deal about how to succeed and fail at siting noxious facilities in the United States and other western nations. The first lesson learned was that early industrial location theory would no longer work. As late as the early 1970s, it was generally assumed that an organization would pick a site, announce, defend, and build it. Sites were picked because they were believed to be the most efficient in regard to access to raw materials, energy, water and transportation. Workers were assumed to be available. Local government would be supportive because manufacturing contributed to growing regional wealth, jobs and taxes. Furthermore, the expectation was that entrepreneurs would spin off new jobs, new projects and tax dollars as a result of adding manufacturing. Waste was deposited somewhere nearby, out of sight and out of mind. With some exceptions, communities wanted manufacturing facilities, and showed their preference by tearing down housing, rebuilding streets, getting rid of waste with relatively little concern. These processes were part of this manufacturing-centric paradigm until the mid-1960s and early 1970s, when that worldview obviously failed the test of reality.

The second lesson learned was that public outrage could force elected officials at local, state and national levels to become concerned about and then change what noxious facilities could do, and where and when they could manufacture transport products and dispose of waste. The National Environmental Policy Act and other laws were passed to reduce emissions into the air, water, onto land, and protect the drinking water supply. A large portion of the U.S. EPA budget was set aside for remediating the worst industrial and urban legacy hazardous waste sites. By the late 1980s, the first solid empirical evidence was produced that poor and certain minority groups disproportionately lived near the most hazardous sites, which raised the political stakes to include civil rights.

By the late 1980s, companies and government managers were pursued by aggressive U.S. EPA regulators and spent millions of dollars to clean up sites. Also, they began to manage waste to minimize negative impact. Some of the first efforts failed, such as using inexpensive landfilling for hazardous waste sites and building ocean going incinerator ships (Chapter 8). Then, they turned toward pollution prevention and continuous safety improvements, which have been much more successful.

The third major lesson learned was at the global scale. Slowly at first and increasing with time, the world economy became internationalized with hot spots and cold areas. Europe and Asia have become major manufacturing centers; China has been the most obvious case, but also South Korea, Japan, Vietnam, and others using the North American and European idea of manufacturing as an economic base. The third lesson was that non-Western nations, not just North America and European centers, were driving the world's economies and the set of economic, environmental, social and political issues that follow from globalization has made siting options more complicated.

Producers and waste managers are still driving siting and non-siting choices, but the obstacles they face have pushed them in four directions. These four directions mark the fourth major lesson learned, which center around the following steps:

- Institute pollution prevention and continuous safety improvement practices to place the organization in a more favorable position with regard to consumer preferences, the legal system, and save resource and associated costs.
- Avoid high risk siting choices in favor of clamping needed facilities to existing sites, as well as redesigning production and products so they are less noxious.
- Engage in negotiations with government and community representatives to assess the likelihood of success in a location and building long-term to support an investment.
- Recognizing local social, political and economic realities means that some production and associated waste management can no longer be managed locally and must be relocated or no longer manufactured.

These four directions are continually tested by new trends. For example, how effectively will these four options be in regard to digital manufacturing, nano and biotechnology siting needs? How well will they respond as sustainability increases?

Adjusting economic thinking and risk analysis

The world economy appears to be moving at an accelerated pace. This brings me to the assert that those of us who were steeped in traditional economic and risk analysis need to rethink how to cope with a rapid motion world economy and a plethora of factors that were not seriously considered in traditional analyses.

Traditional approaches and tools

Traditional economic and risk analysis centered on four tools, although others were also used:

- Cost-benefit analysis;
- Risk assessment;
- Bayesian tools;
- Multi-attribute models.

When these were first applied, the results seemed near miraculous. They gave us a path for finding answers to tough impact and trade-off questions. They allowed us to provide regulators with options that took uncertainty into account. I still use and teach these methods. They work best for the kinds of issues with tightly bounded risks (e.g., cancer is the outcome of a single agent exposure, lead poisoning threatens children). We can use the traditional tools to estimate uncertainty and apply sensitivity analysis and conservative guides to bound options. These tools also work when relatively few parties are involved and they understand the same information.

We continue to need these approaches. We need to know if an agent in commerce could cause birth defects, if a plume from a landfill or a stack emission could cause a health-related problem, and so on. The list of applications of the traditional approaches is increasing not shortening.

Broadening considerations: issues, ethics, and continuous scanning

Siting a noxious facility has become a different species of challenge. Consequences are numerous and debatable. Different ethical bases challenge all stakeholders. Permutations and combinations of important factors in noxious facility siting raise uncertainty to levels where the above four traditional economic and risk tools applied to chemical carcinogenesis, nuclear power plant reactors, hazardous waste cleanup and other assessments are necessary but insufficient. I have tried to apply them, and what I saw was that the tools would have driven the research questions and analysis rather than the questions and potential consequences driving the selection of tools.

In essence, as the complexity has increased, our uncertainty has increased to the point where I doubt we could assume that Bayesian methods and Monte

Carlo modeling would provide impact estimates that would be accepted by decision-makers. Much as I have enjoyed multi-attribute modeling, which allows us to better understand who benefits and does not from alternatives, these classical tools cannot handle an infinite number of attributes, groups and other kinds of variables that occur in real world siting examples.

In noxious facility siting cases, the parties are not going to be satisfied with managers limiting the number of issues. Here is short list of challenges that we need to expect:

- *What are the issues?* (Acute risks? Chronic risks? Human health and safety, environment, cultural impacts, environmental justice, cumulative impacts, economic concerns? Political consequences? Spatial and temporal variations in impacts?) Too often, I and others like me have misjudged what people living in the area think are the most important issues. I have been at meetings at which the public has disagreed with the experts for assuming we knew the issues. At the same meetings, I have seen local groups show distain for outside environmental activists who think they know what concerns the population. Dealing with people who know their home turf is an eye opener.
- *What, if any, cumulative health, environmental, psychological, economic, social and political effects is plausible from each of these risks?* Do these actions produce positive feedback, or negative? For example, I intellectually understood the idea of cumulative impacts, and then I really understood them when I became involved in a siting case in which one group of scientific experts did not take cumulative effects into account and another focused on them. Those who ignored them were castigated and ultimately sued. Today, only the foolish would ignore the idea of cumulative effects. In essence, for siting cases we need to know how a broad range of potential benefits may be enhanced and negatives prevented or minimized.

The challenges that occur in siting noxious require a set of approaches that go beyond the traditional processes. The biggest gap is a lack of universally accepted set of values/ethical principles. Here, I offer a set of six guidelines that I began working on in 1967 when the board of education of the City of Yonkers, New York, voted to take 29 houses, including my parents' house, to clamp on a new school to an old one (the siting proposal failed). Over time, I redrafted these to fit noxious sites. I offer these as a compromise between on the one hand the power of government authority to impose a site on a community and on the other hand giving every local government the power to veto a site.

Here are six ethical guidelines for siting a noxious facility, and Table 10.1 provides probes about using those guides in regard to a proposed site:

1   Some noxious sites are necessary in order to manage a growing world economy, but their size and mix of new, clamping and non-siting alternatives should be subject to deep thought, scanning, and revision.

2   Volunteers sites should be sought, but voluntary sites must pass health, safety and environmental tests in order to be considered.
3   Every effort within reason should be made to use state of the art pollution prevention and continuous health and safety improvements at any proposed new site and the vast majority of clamped ones.
4   More effective guidelines should be prepared by EPA and other agencies to steer sites away from places that already have more than their share of noxious facilities and these guidelines should include consideration of local economic, social and political impacts, as well as human health and environmental effects.
5   Any unavoidable risks must be negotiated, with the highest weights attached to the preferences of the population most likely to be impacted. This means that negotiations may focus on financial, social, economic, human health, and environmental benefits as valued and perceived by the parties.
6   Communities with limited resources should be assisted by government and/or private funders so that they can negotiate with a good chance of having their position adequately represented.

Table 10.1 provides specific questions that should help developers think about what they would need to do to follow these six ethical guides. Some readers may dispute the very nature of some of these questions as outside their domain. However, these questions are highly likely to be asked, and getting ahead of them is probably a good idea.

Following from these six suggestions and the questions in Table 10.1 are some additional issues. One is to determine who can participate and how they can participate. Hansson and Hadorn[1] offer some ethical guides about how to answer these and related ethical questions. Next, even if we know who has standing, we need to be clearer about what roles they play. Do negotiations produce a decision, or recommendations? Should they aim toward a majority, unanimous, or some other level of support? Each case may be different enough to warrant unique answers to these questions.

Methods of listening to different views range from random sample surveys to one-on-one face-to-face interviews. While each situation warrants consideration of all options, I offer the following three expectations about the foreseeable future:

•   Negotiations are trending toward a broader set of representatives. Accordingly, it is desirable to use a committee structure to investigate key issues and then to bring the results to a larger committee with the differences narrowed or characterized as not resolvable.
•   Greater reliance on big data and display of the data with enhanced mapping tools that can overlay data, extrapolate information and allow the negotiating parties to explore scenarios and charrettes that enable them to gain a better understanding of the implications of choices. In some cases these analytical exercises will include interactive modeling to assist the parties.

Table 10.1 Checklist to support six ethical guidelines

---

1. Need for site
What national and international trends, information about current and planned location are the bases of proposing the site?

Is there anything looming in regard to new technology that could make the site obsolete before the investment benefit is realized? New processes? New resource prices? Competitors? New regulations? Markets? New elections in the local area?

Have CLAMP and let-it-go options been evaluated?

Have all non-siting options been evaluated?

2. Volunteers
1. How will volunteers be identified?
2. What information must be known about the volunteer location in regard to the site(s) itself and the local jurisdictions?
3. What was the property used for (title search, aerial photos, fire maps, etc.)? Does it have a regulatory agency permit? (RCRA, CERCLIS, LUST, etc.)? Is there evidence of site disturbance? Who currently owns the site? Are there endangered or threatened species on the site?
4. How will developers evaluate the cost and benefits to the organization, and to the surrounding jurisdictions in regard to local fiscal, regional economic, social, and political impacts?
5. What decision-making process will be used to select the best location(s)?

3. Environmental and ecological health
Will atmospheric discharges from the proposed site, added rail, road or other infrastructure lead to violations of national ambient air quality standards, exacerbate an existing violation? Include releases of toxic substances?

Will the site produce detectable odors?

Will the site produce sounds that exceed 55 dB(A) outdoors or 45 dB(A) inside a residential structure? Will any disturbing sounds be producing during the night hours?

Will traffic in and out of the site disturb local residential neighborhoods? Popular commercial or recreational areas?

Will the appearance of the site disturb local residents?

Will the site disturb local surface or ground water supplies? Change the quantity or quality of water? Will a recreational site(s) be impacted?

Will waste be stored on the site? Will waste be transported from the site by truck, train, and pipeline? If so, how will these activities be managed to prevent releases?

Is the project likely to be located on or near an ecologically sensitive area?

4. Environmental justice
How many people live within ½ mile, 1 mile and 5 miles of site? How has this changed during the last 10, 20 years?

Are the residents disproportionately Afro, Latino, Native Americans or poor? How does this compare to 10, 20 years ago?

What is the proportion of any identified minority or poor population compared to that of the local government (borough, city, county, township) and state 10, 20 years ago?

Are there noxious facilities in the area? What makes them noxious?

Has the proportion of noxious facilities increased in this area during the last 10, 20 years compared to the local government and the state?

Are schools, child-care, homes for senior citizens and disabled, and hospitals within 5 miles? Can you determine how well equipped these sites are to evacuate or shelter in the event of an incident?

Will construction of the project cause the migration of rats, mice, and other unwanted species to adjacent areas? How can this problem be prevented or reduced in scope if it occurs?

5. **Negotiations**

Are there local records of public meetings, focus groups and sample surveys, newspaper stories, media interviews and other records that would help organizations better understand the perceptions, values and especially preferences of legitimate stakeholders? Have there been any negotiations about proposed roads, bridges, malls, and other land uses that provide insights?

Can the organization meet with local elected officials and civic and religious-affiliated organization to talk about perceptions, values and preferences?

Is the project sufficiently compelling that the organization wants to formally invest in sample surveys?

What impact, if any, will the proposed site have on land use? Is it incompatible with any existing land uses? Will it or can it be perceived as threatening property values? Does the site threaten important historical and cultural sites?

Will the proposed site require additional service commitments, such as police, fire, sewer, water and sewer for domestic waste, and garbage collection? How would these be funded?

What fiscal impacts will this proposed site have? Will it add local jobs? Local taxes through more employees and local businesses? Is a local job-training program needed for potential employees?

What are likely impacts on health, safety, and environmental protection? (See #3 above)

How will the developer respond to challenges in regard to each of the above areas?

6. **Technical assistance**

Based on local history, including letters written as part of EISs, media reports, local presence of universities, and other expertise, how prepared are local parties to negotiate the proposed site?

Assuming that gaps exist in local capability, is the developer willing to support independent experts to fill this gap that are approved by the local group and community groups?

- Continue use of public meetings, however, there should be much less reliance on them for substantive outcomes because of their tendency to degrade into rhetorical exchanges that encourage some to dig their heels in and vent rather than lead to productive speaking, listening, and negotiating.

Given that I expect us to have more data, and enhanced simulation and mapping tools than we have had, there is a need for more thinking about bounding problems to aid decision-makers with weighing evidence. I advise the following four efforts:

- Provide information to users and audiences about the use of symbols, values and other mapped information and about the limitations of simulation models.
- Provide guidelines about how to weigh evidence.
- Build decision-informing heuristics for negotiating parties to use, debate, and modify.
- Prepare guidelines for using sensitivity analysis.

More involvement but limiting data access is a problem. We are now in a period where technology allows us to share more information, and yet data sharing is often discouraged and not permitted because of trade secrets and national security. The references listed at the end of Chapter 9 provide clues about where noxious facilities will be needed. However, much more detailed analyses exist but are not publically available. The net effect is to make it difficult for interested parties to fully engage siting-related issues with the same preparation. Network hacking has only complicated this dilemma. These emerging trends are a problem, and acting on solutions that will provide better access is imperative in order to avoid conflict caused by information stove-piping, in other words, a dispute over a site attributable to values and perceptions is expectable, but parties should not argue over known facts that are not available to some participants.

An exception to this sour trend is efforts by government agencies like the U.S. EPA to collect more data and share it. In the area of environmental health, the chances for scanning, ongoing surveillance, adaptive management, and resilience aided by enhanced computer and modeling capacity is increasing.[2] Their goal is to depict cause and effect relationships or at least find key influences on human health and safety. It is the economic side of the siting issue that has become more challenging because of trade secrets and security issues. I do expect more scanning as follows:

- Scanning local, state, national and international data sets looking for trends that will help predict the need for siting and non-siting solutions;
- Scanning environmental health data sets to look for improvements in sustainably, resilience, pollution prevention, and health and safety; and
- Scanning social, economic and political data sets to understand trends that could impact siting choices.

The ideas expressed in this section are not new. The limitations of traditional economic and risk analysis in regard to gathering, depicting, and using value and ethical-based data are challenging, and the need to scan for new trends are established in the literature, albeit not firmly in the noxious facility siting literatures.[3-5]

Coping with chronic garbage-related-challenges

The noxious facility types described in Chapter 1 have an interesting history and a challenging future. Chemical plants, oil refineries and gas plants, metal producers, paper and animal product facilities face stiffer opposition in many areas than they did decades ago. But they appear to have found places that not only tolerate them, but have built their economies and communities around them. Furthermore, long distance transportation capacity provides options that did not exist in the twentieth century. These noxious facilities will manage the siting challenge, even if they never come close to finding the most economically efficient sites or the least human health and environmental-health-damaging locations.

Buildings product sites have been adopting much stricter particle emission controls and by-an-large remain in their market areas because of the high cost of moving the products. With regard to energy production, coal plants and some nuclear power plants are being shut down in the United States, but not everywhere. Some of shuttering has to do with technological obsolesce, public concern about these facilities, and in the United States much of it has to do with the current availability of cheap shale gas. Each of the types of noxious facilities in this and the prior paragraph depends upon a market, and hence we should expect that their managers will develop ways of coping with changing markets and expectations for performance, including health and safety, sustainability and the others described in this book. I am not saying that finding new sites is going to be easy. I am saying that there are options, especially if developers can provide pollution prevention and continuous safety upgrades, can use clamping and non-siting options, are willing to negotiate, and recognize that they may need a variety of coping strategies because some objectives may take longer to achieve than others.

Waste management facilities are at the top of noxious facility list in Chapter 1. Given the national and international attention focused on nuclear defense and power plant waste, arguably the nuclear industry has the most coping to do. But I would not rate it as the most challenging because the reality is that both the defense waste and nuclear power plant industry have default backup plans. High-level defense waste can stay at the nuclear sites for many decades and the power plant waste can stay at the sites where the waste is produced. Neither option is a great choice from a health and safety perspective, and legally not acceptable given past agreements. But the sites where the waste is currently stored are large, and the companies and government are willing to spend a great deal of money on security and monitoring the sites. Hence, despite the ongoing controversies over storing nuclear waste, current nuclear waste management practice in the United States does not represent an immediate emergency. Low-level waste has similar options that require transportation to licensed sites.

Chemical waste management facilities remain a serious challenge, but progressive leaders in this industry are targeting the reduction and reuse of waste. Some sites will need to be added, and these will need intensive monitoring to prevent dangerous emissions. Older sites will be closed. But this industry seems to be heading in the direction that makes sense from economic and risk perspectives, and I see it as more tractable than the final type on my noxious facilities list.

This leaves me with household garbage facility siting in large metropolises as the chronic noxious facility type with the most need for coping, or perhaps the right label is muddling through. In 2016, Adler[6] reviewed studies that rank wasteful cities, that is, cities that produce a great deal of waste. New York City is at the top of the list because it produces so much trash per person and per acre. He then discusses Houston, Atlanta, Tampa, and Phoenix, which are sprawling cities that have worse environmental metrics than New York City, but in regard to siting have more nearby disposal options than New York City.

Overall, in the United States and other dense North American and large European cities, in regard to trash management the following four problematic conditions exist:

- Public and commercial waste generation continue to increase and the waste must be picked up to avoid human health, safety and environmental problems at rapidly increasing cost.
- Many large urban centers in North America and Europe have already used up nearly all or all of the nearby space where waste could be burned or buried.
- Historically, household garbage contains hazardous materials and has been mixed with hazardous commercial and industrial waste (Chapter 3); so-called garbage or trash has many dangerous components.
- With few exceptions, communities do not want a waste site bringing odors, truck traffic, unsightly appearance, bird and rat populations, and potential for fires, air and water contamination near them.

New York City: metropolis central

The historical siting approach

Whether New York City is the most wasteful city in the world is debatable. But even if it is not on top, it certainly is in the discussion. I am familiar with its coping mechanisms, and hence I summarize New York City's challenge, focusing on siting and non-siting mechanisms.

The City of New York has gone through a "revolutionary-like" change in its solid waste management approach in the twenty-first century. New York City has had over seven million residents for over 70 years, and now it has over eight million. For much of the last century, it had a solid waste management strategy of finding multiple landfill sites, filling them up, and moving on. But this policy stopped working because the sites were consumed. In 1945, my parents took me, then less than two years old, to a farming area in Staten Island, New York City, where they purchased fresh vegetables. Two years later, the salt marsh area adjacent to farming area was chosen as a "temporary" location to increase New York City's garbage handling capacity. This temporary site was needed because sites located in all five counties (called boroughs) were full.

The reasons for choosing Fresh Kills on Staten Island are uncertain, but several reasons are easy to point to. In 1940, Staten Island had a population of 174,000, about 2% of the New York City population. The area was rural, cutoff from the remainder of New York City by water bodies, that is, there was no subway or elevated train to bring massive numbers of commuters to Staten Island. My family travelled there by ferry from lower Manhattan to buy vegetables, and my father had gone swimming on beaches in Staten Island when he was a boy, which would have been in the 1920s (see below for Text box 10.1).

Text box 10.1 Father, mother and uncle interview about Fresh Kills closing, notes from July 5, 2002

Sydney Greenberg, my father, operated a printing business in lower Manhattan:

I suppose that I should be pleased that they finally are trying to do the right thing. But they will never get it back to what it was. Staten Island was a beautiful place that we went to for fun. My dad [Phil Greenberg], mom [Sophie], and sister [Beatrice] would take the subway from the Bronx to lower Manhattan, and take the Staten Island Ferry to the beach and go to the Fresh Kills area. I would swim, go down to the stream and shoot rocks with my slingshot. Michael, they killed that place. The park they are building is better is better than the mess they left, but it doesn't make up for what they did. They should have built incinerators and burned the trash and recycled a lot of the reusable material, which is what we did during the war [World War II]. I recycled so much of the chemicals, paper and plates in my business. It took some work, but it was what we did. Now I see the same stupid mistakes being made in Mexico, China, and India with these mountains of garbage [see below]. It is sad.

Sol Saletra, my uncle, was a senior official in the U.S. Immigration Service stationed in New York City:

Syd's right. But honestly, I think Fresh Kills was a way they could make sure that the underworld could maintain control of the garbage business in the City. To me it was about graft. Also, my co-workers would tell me that the Mafia buried people in Fresh Kills.

Mildred Greenberg, mother, an accountant for many years: Syd is right about how sad it was. But unless you visited Fresh Kills, it was out of sight, out of mind. No one really cared about garbage. No one really focused on garbage until the garbage strike, and then the dopey cops gave Grandma Rose a ticket [see above]. A lot of the garbage problems go back to Mayor Lindsay who was a handsome guy, but really had no idea about people. He wanted to keep the unions down, and Mike Quill [head of the transport union] was jailed during the 1966 transit strike. Quill told the judge to "drop dead" and called Lindsay by the name Linsley. This is how it worked then. People ignored the environment until it kicked them. They focused on today and left the future for the next group. My generation was all about getting a house, a refrigerator, a good stove and a TV, not about your generation's environment.

Mexico City, Calcutta and all these places you talk about, they really are poor. How are they going to manage garbage when they can't feed or educate people? Here, in New York, we didn't have a poverty excuse. Our values, priorities were different.

Idyllic Staten Island was vulnerable. The comparative isolation of Staten Island meant that 20 barges a day (about 13,000 tons of garbage) from all the New York City counties could unload the waste and space was available. That part of Staten Island became the Fresh Kills landfill and the entire massive site was covered by garbage.

The massive 2220-acre landfill began to rise and become as visible a symbol as the Empire State Building and World Trade Center. In 2001, Staten Island's population was 450,000, and county officials were fighting New York City over the stigmatizing impact of the massive structure. The landfill was temporarily closed. It reopened only after the 9/11 attacks on the World Trade Center, and a good deal of the rubble was buried there, and police searched through the rubble to identify some of those killed during the event. In 2014, the world's formerly largest landfill was closed; a park is being built and is expected to be expanded over the next three decades (Figure 10.1).

A new coping strategy

New York City knew that it needed a new strategy even before Fresh Kills was closed. In 2001, their written strategy[7] included all of the major options described in Chapters 5 to 9:

- *Pollution prevention*: The plan emphasizes recycling and use of containers to pick up and transfer the waste to transfer stations and then ship it by barge and train. This plan takes a formerly loosely connected set of employees, garages, vehicles, private haulers, transfer stations, and disposal companies,

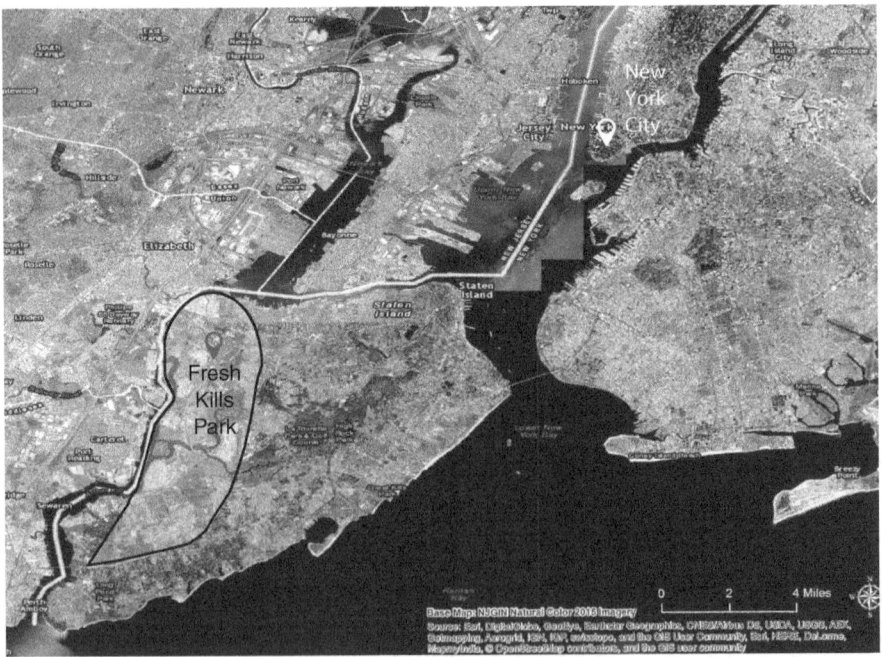

Figure 10.1 Fresh Kills park.

and weaves them into a more coherent system to more effectively and efficiently manage the waste, reduce costs, and notably reduce exposures. The
number of garbage trucks with their noise, air pollution, energy use and
road hazard potential is minimized (many are still in use on local roads),
and the trucks are designed to use less polluting fuel.

- *Clamping*: A major goal is to identify a limited number of strategically
  located sites in each borough for recycling and trash transportation. This is
  accomplished by identifying transfer stations on city-owned land, and
  relying on barges and trains to ship waste from these sites.
- *Negotiation*: The plan no longer places the garbage center label on Staten
  Island. The administrative structure includes representatives from every
  borough to devise local plans, and as much as possible divides responsibility
  among the five boroughs. The engagement process also includes public and
  private participants, environmental advocates and those that have knowledge and a stake in the system.
- *Let it go*. New York City could not design a politically acceptable way of
  managing all the garbage in the City. Hence, it signed long-term contracts
  to ship it by rail out of state.

The more I studied the plan and spoke with people, the more it represented a
text book coping response to a siting decision with no local sites. What is also
clear is that New York City is also not stopping with the current plan. New
York City is currently scanning other opportunities and notes that the current
system is "unsustainable in the long run"[7] (p. ES-1). It has a fund for pilot
studies, including building pilot plants to test out technical options, for example
for recycling plastic and paper. The report claims that New York City is already
a national leader in using alternative fuels in its fleet of garbage trucks, including bio, fuel cells, propane, ethanol, methanol and hybrid fuel trucks with electric. I have seen several of its efforts to explore compositing, anaerobic digestion,
and other options for managing selected waste. New York City has begun an
electronics recycling program, and is working on odor control at legacy sites and
recycling facilities. A long-term goal of the plan is to contribute no waste to
landfills, which seems like a dream right now.

The current NYC plan stretches out to the year 2025. I do not like the fact
that New York City is shipping garbage hundreds of miles to Pennsylvania, Virginia, and other states. Nor do I believe that New York City officials are comfortable with all the choices they have made. The policy contradicts the idea
that waste should not be created, it should be recycled, and then if necessary
locally managed. But New York's reality is that it could not completely manage
the garbage locally. They needed to cope by designing a system that will take
them forward until a new system could be designed. Not to develop a plan that
could be implemented courted a serious economic, political, and human health
problem.

In regard to the human health consequences, I lived in New York City's
"great garbage strike of 1968." It was one of many strikes in that period that

included taxi, tugboat, public school and even grave diggers.[8] The strike began on February 2 and featured angry words, clashes between the workers and police. It was ugly.

I could see then that this strike would soon lead to a public health emergency that was a lot uglier than name calling. After day three, I was trying to figure out how to evacuate my grandmother from her tiny one-bedroom New York City apartment. For years she had deposited her garbage in large enclosed bins and the building superintendent brought these bins to the curb where they were collected by sanitation workers. During the strike, these receptacles were overfilled with trash, and so she and others began leaving it out in the street. In my grandmother's case, a policeman gave her a summons for littering. Hence, I needed to get her out of there. In fact, there was no pickup for nine days, and the odors and rodents were taking over the space and there was a fire threat. The mayor at the time and the unions at the time disliked each other and both seemed prepare to jeopardize the health of many people to win a political duel. Fortunately, Nelson Rockefeller, Governor of New York State, intervened using a public health emergency rationale.

Now, New York City has an enormous advantage of wealth and experience in managing waste. The New York City Department of Sanitation (DSNY) has 9700 employees, an annual budget of $1.4 billion, disposes of 3.2 million tons of refuse and more than 600,000 tons of recyclables a year.[7,9] The Commissioner, a native New Yorker from Brooklyn, interned in the DSNY and appears committed to a modern sanitation service, rather than the one I recall as a boy.[9] Underscoring my confidence that New York City can overcome its garbage legacy and move toward a more efficient and less destructive system is a letter of approval for the New York City plan by Carl Johnson, Deputy Commissioner of the New York State Department of Environmental Conservation who had the following to say about the plan:

> The plan reinforces the State's commitment to sustaining and managing our resources, environment and economic competitiveness by placing emphasis on waste reduction and recycling, while providing an equitable waste management infrastructure where the needs of its residents, businesses and industry are met.[7]
>
> (p. 1)

Tokyo and Singapore represent other massive cities that have chosen a different path than New York City but have the assets to find a solution. I would advise reading Tokyo's.[10]

How is implementation of the current plan progressing? In 2001, 20% of New York's household waste was recycled.[7,11] The proportion recycled dropped to 11.5% in 2003, but it rose again to 15% by 2005. Fiscal year 2017 data show a rate of 19.2% for curbside and 17.3% for DSNY managed recycling. New York's plan called for more recycling, but it is difficult to implement the plan because so many New York City residents live in apartments. It is difficult to monitor their recycling. The lowest recycling rate in 2017 was in the Bronx – 12.9% – and the highest in Staten Island – 22%. The former is relatively poor and had only 19% living in single family homes, whereas the second is more middle class and over two-thirds lived in single family homes.

The changeover from truck to train has begun to work. Cohen[12] reports that 32% of the trash is transported out of NYC by train, 23% by regular sanitation truck, and 45% by long-haul truck. The proportion by rail and barge is eventually expected to increase to 88%. NYC has locations where barges can pick-up the waste and transport it out of state. To date, these out of state locations have included Ohio, Pennsylvania, Virginia, and South Carolina, which are hundreds of miles away.[13]

Realistically, in the absence of waste-to-energy facilities, which certainly qualify as LULUs, a massive step up and recycling, and new product designs that create less garbage, the temporary solution for urban centers is long-distance shipping out of the area where garbage can be a revenue source. However, it is only a matter of time before opposition to the long-distance shipping practice accelerates and gains political force. At this time, however, in the U.S. garbage is legally viewed as interstate commerce and accordingly cannot be arbitrarily stopped by out-of-state officials. Hence, as long as New York City can pay, it can ship trash all over the country. It remains to be seen how long this shipping out approach to municipal garbage collection will be tolerated. New York City is aware of this reality and is looking for non-siting solutions and clamping rather than large new facilities and long-term shipment.

I looked for a cumulative assessment about New York City's response to garbage management over the century. By far the best I found was in notes from a meeting on July 5, 2002 with three lifelong New Yorkers: my father, mother, and uncle, who also had taken notice of what they saw as other cities repeating the same mistakes.

I am much less sanguine about large cities in less developed nations that have far fewer resources and experience. The growth of their populations and economies has placed unprecedented demands not only on waste management, but roads, education, health care and every other service. These include Mexico City, Rio de Janeiro, Cairo, a number of major Chinese cities, Manila, Jakarta, and New Delhi, all of which are making progress, but have enormous legacy and future garbage-related problems to face.[14,15]

New Delhi: a rapidly developing metropolis

The world's large cities provide ample opportunities to find a place that is trying to cope with much more serious garbage problems than New York City and without the resources of New York City and the history of building a system that picked up and deposited garbage.

New Delhi, the capital of India, is illustrative of trying to use modern approaches under enormous pressure. In 2014, the population of India was estimated to be 1.27 billion, and the population of New Delhi was 17.8 million and growing rapidly. For years all the city's waste went to three landfills that were more than twice the size of their designs. Pandey[16] reported that its three landfills were full and that the city was facing a garbage crisis. The author reported that 500 more truckloads every day of garbage were being brought to the three sites, and yet more garbage was arriving. The article quoted business interests as saying that garbage at malls and restaurants was piling up. Photos and additional articles show it to be far worse than even Pandey's distressing article suggests. Waste simply flowed from the sites polluting the water to a color described as the color of Coca-Cola. That water seeped into water bodies and groundwater and became local drinking water. On the surface of the site, without proper compaction, pockets of methane formed and periodically fires would break out, not be extinguishable, and pollute the air with unpleasant odors and toxins. I have seen and described this outcome in Chapter 3 at the Kin-Buc Landfill, but without having visited these New Delhi sites, they seem an order of magnitude worse, which, frankly, is hard to believe.

Nandi[17] characterized the three landfills in the New Delhi area as follows:

> These are monstrous trash mountains, including hazardous waste, leaching out toxic liquids and emanating noxious fumes. Thousands of scavenging birds swarm over them as they grow larger every day.[17]
>
> (p. 1)

The author then discusses the landfills as a source of jobs for local people (also see[18]) and criticizes management for not recycling demolition waste, bricks, concrete, wood, and rubble, or recycling organic waste. The article continues by focusing on heavy metal contamination found offsite and fires at the sites. There is no monitoring, and leachate seeps through the landfill, and some workers typically contract asthma and are bitten by stray dogs.

Annepu[19] proposed a strategy that emphasizes recycling, biomethanation, biological treatment, waste to energy and refuse-derived fuel that he asserts would remove over 90% of the waste going to landfills. The pressure to find sites for landfills, and build waste-to-energy plants (WTEF) in India as a whole and New Delhi, in particular, is formidable because of rapid population growth and lack of space and infrastructure to move the waste, and the presence of numerous other priorities to absorb government and private resources.

Writing in 2014, Zerkel[20] described the metro area of 21 million as in the middle of a trash crisis because the city's solid waste production increased 50% in five years and the landfills cannot handle the waste. The vast majority of residents have no trash pick-up and 350,000 residents of New Delhi pick over the trash as their job.

I could go on, but there is nothing subtle about the seriousness of the problem described in the literature. The capital region of India had no system in place and needed to use many of the tools described in this book to build a functional one that included all the areas. There was no complete package to pick up all the waste, no clearly demarcated and acceptable places to deliver the waste, no demonstrated method that would not notably create human health and safety and environmental damage, and places that would recycle the waste.

In 2001, when New York City had to redesign its system to recognize the lack of local disposal sites, the Delhi region also began to acknowledge that it needed a plan to manage garbage. The regional government[21–27] and other parties acknowledged that

- Solid waste management was managed piecemeal rather than as a interconnected system;
- A plan was needed to pick up the garbage and workforce was needed to accomplish that task;
- Decision-makers did not understand the problem or what needed to be done;
- Funds were lacking for management and planning;
- The public did not understand the problems; and
- Land for new sites was limited.

Its authors expect waste generation to double between 2001 and 2021, and they recommended the following:

- A solid waste management plan;
- Standards for collecting, transferring, and disposing of garbage in less environmentally destructive ways;
- Identification of sites for waste management and then earmarking them for this purpose, which is a challenge because many areas repeatedly flood;
- Creation of recycling and waste minimization programs;
- Development of a public education program through the media; and
- Changes in the institutional processes of managing the waste to focus on efficiency and risk reduction and involve stakeholders in this process.

I do not have any personal contact with staff working on the Delhi region's solid waste case. Using reports, it seems clear that three of the four approaches described in Chapters 5 to 8 are in play.

- *Pollution prevention:* The plan lays out a process for collecting all the waste, they have banned the burning of plastics, created new rules for managing e-waste, a construction demolition waste plant has been built, and a new bio-remediation facility is underway. Most notably, the area's largest WTEF facility opened in March 2017. It burns 2000 tons of garbage and creates 24 megawatts of energy.[28]
- *Clamping:* With the caveat that I have not found a map of all the new facilities, they appear to be sited next to or on existing waste sites.
- *Negotiation:* A major strike shut down garbage collection in the area. Negotiations settled the worker garbage strike dispute and like the New York City case described above galvanized efforts to build a manageable system. Also, private and picker jobs were part of the negotiations to build a collection system for the entire area.

- *Let it go*: This region appears to be focusing on what it can do locally, and following the model of Singapore, which heavily invested in WTEF and recycling after its waste increase almost seven times between 1970 and 2016.[29]

The Delhi region has enjoyed economic success and is challenged by social, environmental and political issues. Progress has been made in managing the massive garbage problem. If this continues to be priority, siting and no-siting options as part of implementable plans should become more efficient and protective. I have no ability to say the extent to which the changes in economic and risk-related tools and processes described in the chapter have been used, but it is hard to believe that they would have gotten this far without using many of them. Overall, Delhi's coping is more of a struggle than its large Western counterparts face. I would say that it is muddling through.

Coping with an opportunity from the past

Although many of us would like to return the world back to the time when they thought it was best for them, the world was probably not so wonderful for many people. I decided that the last case study should be a real shocker. The movie the *Day After Tomorrow* begins with the premise that humans have abused the Earth and caused a new ice age in the northern hemisphere. The movie concludes with the U.S. President and other residents of America moving to South America. Whether they would be welcomed is not discussed, and frankly I doubt they would be welcomed.

I decided to create a hypothetical return to the past policy that would come as shock to many planners and decision-makers working in Western nations. My return to the past was to be a local or state government that decided that it wanted to allow industrialization along their coast. Then I was going to play out what might happen when the policy idea was proposed based on the ideas presented in this book.

To my complete surprise, a month before I began writing this chapter, a policy proposal introduced in the State of Delaware came close enough to my hypothetical for me to use the actual case as a demonstration. For context, Delaware is a small state (49th out of 50 in land area). The state is much longer north to south than it is wide (30 miles wide and 96 miles long [48 km by 154 km]), which means that it has a long coast line.

The state lies in the middle of the northeast United States urban-industrial corridor between Boston and the Washington DC metropolises. It borders New Jersey, Pennsylvania, Maryland, and Virginia. With regard to demographics, Delaware's population is about 950,000 and the state has the sixth highest population density (over 470 per square mile). Like its neighbors, the state's population has been increasing in regard to Latino and African Americans who constitute over one-third of the population. Also like its neighbors, Delaware has hemorrhaged manufacturing industries and more recently other companies.

For example, DuPont has been headquartered in Delaware. Then it merged with Dow, and there is great fear that nearly all of its signature manufacturing industry will eventually leave. General Motors and Chrysler closed assembly plants, MNBA was bought out. Government officials are rightfully concerned.

Before these events occurred, in 1971, Delaware passed the Coastal Zone Act (Title 7, chapter 70, Delaware code) that made it extremely difficult to locate an industrial facility within two miles of the 115-mile long Atlantic Coast and inland bays.[30] Fourteen sites that already existed were allowed to continue. The law also stopped development of a deepwater port. The Delaware law became a blueprint for other areas across the United States. Many interpreted this law as the state choosing recreation use of coastal water over industrial. Legal cases were brought against the law, but were unsuccessful.

Forty-six years later, in May 2017, two state legislators introduced a bill that would allow new heavy industry to locate at the 14 grandfathered industrial sites, and bulk product transfers operations were permitted at these sites. In essence, this law would permit clamping at 14 locations, or about 2% of the length of the coastline.

The bill brought out expected polarized positions. Proponents, including the chamber of commerce, other business organizations and individual companies, argued for the following advantages:[31,32]

- Create construction and production jobs, tax revenues, and personal income;
- Rebalance land use along the coast, which they assert is out of balance in favor of recreation;
- New industrial facilities would bring pollution prevention and health and safety practices and thereby be less destructive of human health and the environment;
- Crude oil and refineries, pulp and paper plants, incinerators would not be allowed and only 2% of coastal land is involved; and
- New industries would clean up existing polluted abandoned sites (5 of the 14 are no longer used).

The Bill, HP190, was opposed by more than two dozen organizations representing environmental protection interests including the League of Women Voters. The asserted the following disadvantages:[33,34]

- Increase air and water emissions;
- Impact nearby wetlands;
- Hurt the tourist industry;
- Increase the chances of rail accidents and coastal spills;
- Lead to environmental justice impacts;
- The state of Delaware's environmental organization does not have resources to manage the program;
- The public was left out of the process; and

- The law would fail because the demand for noxious facility sites has been declining in the United States.

On June 7, 2017, Delaware's House Natural Resources Committee voted 9 to 1 to advance the bill to the full legislature.[35] The governor of Delaware signed the bill into law as an amendment Title 7 of the Coastal Zone Act on August 2, 2017. Stay tuned for more about this case in Delaware.

The Delaware case has already brought out the polarized positions that mark discourse about noxious sites. I read the proposal with the six ethical guidelines presented earlier in the chapter in mind:

- Some noxious sites are necessary in order to manage a growing world economy, but their size and mix of new, clamping and non-siting alternatives should be subject to review.

  Comments: the bill limits the type of noxious facilities. It is not clear what the response would be if another facility type that is neither allowed or prohibited were to apply, for example, technologies that did not exist in their current forms when the sites were built, such as biotechnology and nanotechnology. I mention these because there may be few opportunities to open the sites to the kinds of facilities permitted.
- Volunteers should be sought, but voluntary sites must pass health, safety and environmental tests in order to be considered;

  Comments: The devil is in the details, but every site will have different conditions and only unbiased investigations can determine what is relatively low risk and high benefit for each site.
- Every technical and financial effort within reason should be made to use state of the art pollution prevention and continuous health and safety improvements at any proposed new site and the vast majority of clamped ones.

  Comments: It seems clear that this is what is anticipated by the proponents of the idea. But is there sufficient expertise to verify that such a promise will be carried out? How much will it cost to comply with my guideline?
- More effective guidelines should be prepared by EPA and other agencies than currently exist to steer sites away from places that already have more than their share of noxious facilities and these guidelines should consist of local economic impacts, as well as social and political impacts.

  Comments: As a reader of many EIS documents, I have found the guidelines inadequate for environmental justice and economics compared to biological, chemical, and other environmental impacts. I realize that this element was retrofitted to the EIS process, but having inadequate economic impact and environmental justice guidelines tilts the information base away from what might be vital to some parties.
- Any unavoidable risks must be negotiated, with the highest weights attached to the preferences of the population most likely to be impacted.

This means that negotiations may focus on financial compensations, and/or other forms of benefits.

Comments: The intent of the proponents and the chamber of commerce groups are indicated, as are those of the opponents of any change. They offer the polar stereotyped responses. I am not clear that public surveys have inquired about local public's preferences are, nor do I know if any conversations with local official have taken place to know what are their priorities. Jobs and taxes sound reasonable, but there may be other preferences.

- Communities with limited resources should be assisted by government and/ or private funders so that they can negotiate with a good chance of having their position adequately represented.

Comments: I have not seen anything about this element, but as someone who has played the role as technical support for communities, I expect that they would appreciate the assistance.

I would expect some form of cost-benefit analysis would be used to assess proposals. It is not clear how risk assessment, Bayesian tools, and multi-attribute models fit. Where they logically fit is if specific hazards are of concern, these would be evaluated with traditional risk assessment methods, for example plume models to assess the impact of chronic and acute discharges. Multi-attributes models could be applied to each site, but make sense for the full set of sites as a whole, that is, across a set of 14 locations multi-attribute tools could be used to estimate the impacts across the full set of sites or at least a subset.

I would also expect some negotiations with developers, the state government, local government, and representative local groups. Prudence suggests that parties not rely on public meetings to gauge public reaction. Appointing a committee representing parties with a real stake in the outcome would allow a full committee and sub-committees to identify key issues and identify points of agreement and disagreement. As part of these interactions every opportunity should be taken to use tools identified earlier in the chapter to identify, examine, display, and debate the consequences of redevelopment options.

In the long run, this radical policy change may be a bad idea. But without looking at the options it is hard to know if this is a good idea for any of these areas. Responses based on unchangeable values and preferences allow no room for talking, thinking, analyzing and reaching reasoned decisions.

Final thoughts

Every noxious facility siting case is not a contest between good and evil. I am not about to say that greed does not hold sway in some places of the world in the early twenty-first century. Nor will I stipulate that every economic growth proposal should be rejected. We need to get past simplistic polarized characterizations of options, organizations, and individuals. We have amassed sufficient knowledge and have tools that allow us to make imperfect but more informed decisions than in the past. I have worked on cases where people's values and perceptions started far apart, but when they began to work together, trust increased and they arrived at a place where they reached decisions that they could accept. I admit to believing in democratic process, albeit it has its warts. What I have proposed here is that more effort be devoted to using what we have learned and can learn rather than shutting the door on different goals, knowledge, values, perceptions and preferences.

References

1 Hansson S, Hadorn G, eds. *The Argumentative Turn in Policy Analysis: Reasoning about Uncertainty*. New York, Springer, 2016.
2 Gwinn M, Axelrad D, Bahadori T, Bussard D, Cascio W, Deener K, Dix D, Thomas R, Kavlock R, Burke T. A Public Health Perspective on 21st Century Risk Assessment. *Am J Public Health*. 107, 7, 1032–1039, 2017.
3 Etzioni A. Mixed-Scanning: A "Third" Approach to Decision-Making. *Public Administration Review*. 27, 5, 385–392, 1967.
4 Stirling A. Science, Precaution, and the Politics of Technological Risk. Converging Implications in Evolutionary and Social Scientific Perspectives. *Annals of the NY Academy Science*. 1128, 95–110, 2008.
5 Stirling A, Scoones J. From Risk Assessment to Knowledge Mapping: Science, Precaution, and Participation in Disease Ecology. *Ecology and Society*. 13, 14–33, 2009.
6 Adler B. Which is the World's Most Wasteful City? *Guardian*. www.theguardian.com/cities/2016/oct/27/which-is-the-worlds-most-wasteful-city. Accessed March 27, 2018.
7 DSNY. Comprehensive Solid Waste Management Plan. The City of New York. Department of Sanitation. www1.nyc.gov/assets/dsny/about/laws/solid-waste-managementpplant.shtml. Accessed July 9, 2017.
8 Marton J. Today in NYC History: The Great Garbage Strike of 1968. http://untappedcities.com/2015/02/11/today-in-nyc-hsitory-the-great-garbage-strike-of-1968. Accessed July 9, 2017.
9 Leadership. Kathryn Garcia Commissioner. www1.nyc/gov/assets/dsny/about/inside-dsny/leadership.html. Accessed July 9, 2017.
10 Ministry of the Environment. Solid Waste Management and Recycling Technology of Japan: Toward a Sustainable Society. www.env.go.jp/en/recycle/smcs/attach/swmrt.pdf. Accessed July 18, 2017.
11 New York City Department of Sanitation. 2017 Fiscal Year-to-Date Report. New York City Municipal Refuse and Recycling Statistics. www1.nyc.gov/assets/dsny/docs/about-dsany-collections-FYTD.pdf. Accessed July 10, 2017.
12 Cohen S. 2012 NYC Takes out the Garbage. Huff Post. www.huffingtonpost.com/steven-cohen/nyc-takes-the-garbage-out_b_1210334.html. Accessed March 27, 2018.
13 Florio A. New York City Trash: Where Does It All Go? Sierra Club New York City Group. http://nyc.sierraclub.org/2022/08/new-york-city-trash-where-does-it-all-go. Accessed July 23, 2014.
14 Zafar S. Garbage Woes in Cairo. *EcoMena*. November 2, 2016. www.ecomena.org/garbage-cairo/. Accessed March 27, 2018.
15 Godoy E. The Waste Mountain Engulfing Mexico City. *Guardian*. www.theguardian.com/environment/2012/jan/09/waste-mountain-mexico-city. Accessed March 27, 2018.
16 Pandey N. Landfills full, Delhi Stares at Garbage Crisis. *Hindustan Times*. January 14, 2014. www.hindustantimes.com/delhi-news/landfills-full-delhi-stares-at-garbage-crisis/story-wpqivD5DYon50OaO5vxZ5M.html. Accessed March 27, 2018.
17 Nandi J. Capital dumps a fortune at its landfills. *The Times of India*. 2014. https://timesofindia.indiatimes.com/city/delhi/Capital-dumps-a-fortune-at-its-landfills/articleshow/34131703.cms. Accessed March 27, 2018.

18 Sarkar P. Solid Waste Management in Delhi – A Social Vulnerability Study. In M Bunch, VM Suresh, and TV Kumaran, eds. Proceedings of the Third International Conference on Environment and Health, Chennai, India. 15–17, December 2003, Chennai: Department of Geography, University of Madras and Faculty of Environmental Studies, York University, 451–464. www.seas.columbia.edu/earth/wtert/sofos/ Sarkar_SWM%20in%20Delhi%20-%20A%20Social%20Vulnerability%20Study. pdf. Accessed March 27, 2018.

19 Annepu R. Sustainable Solid Waste Management in India. Master of Science thesis, Columbia University, New York. 2012. www.academia.edu/2077298/sustainable-solid_waste-management_in_India. Accessed July 21, 2014.

20 Zerkel E. Trash Cities: the World's Worst Garbage Problems. https://weather.com/science/ environment/news/worst-cites-trash-garbage-problems-20139… Accessed July 12, 2017.

21 Department of Environment, Government of NCT of Delhi. Waste Management. 2016. www.delhi.gov.in/wps/wcm/connect/environment/Environment/Home/Environmental+ Issues/Waste+Management. Accessed March 27, 2018.

22 National Capital Region Planning Board, Ministry of Urban Development, Government of India. Solid Waste Management Sector in the National Capital Region. 2017. http://ncrpb.nic.in/solid_waste.php. Accessed March 27, 2018.

23 Moy P, Krishnan N, Ulloa P, Cohen S, Brandt-Raul, P. Options for Management of Municipal Solid Waste in New York City: A Preliminary Comparison of Health Risks and Policy Implications. *Journal of Environmental Management*. 87, 73–79, 2008.

24 Kaushik P. MCD Polls 2017. BTP Plans Two-Tier System to Clean Garbage in City. April 10, 2017. http://indianexpress.com/article/cities/delhi/mcd-polls-2017-bjp-plans-two-tier-system-to-clean-garbage-in-city-4606872/. Accessed March 27, 2018.

25 Municipal Corporation of Delhi (2010) Landfill Biogas Project Opportunity, Gazipur Landfill, New Delhi, India. www.globalmethan.org/Data/LF_In_Gazipur_flyer_2010. pdf. Accessed July 12, 2017.

26 Ghose D. Waste Management is Imperative in Delhi as the National Capital Inches Closer to Another Deonar. FirstPost. July 13, 2017. www.hindustantimes.com/delhi-news/municipal-corporation-inaugurates-india-s-largest-solid-waste-to-energy-plant-at-narela/story-dZuZaGLV3UFQPzU8vmSbyM.html. Accessed March 27, 2018.

27 Sharma V. Garbage Crisis in East Delhi: Workers on Indefinite Strike, Waste Piles Up. *Hindustantimes*. January 7, 2017. https://timesofindia.indiatimes.com/city/delhi/ east-delhi-sanitation-workers-go-on-indefinite-strike/articleshow/61047489.cms. Accessed March 27, 2018.

28 Sharma V. India's Largest Solid Waste-to-Energy Plant Launched at Delhi's Narela. *Hindustantimes*. March 10, 2017. www.hindustantimes.com/delhi-news/municipal-corporation-inaugurates-india-s-largest-solid-waste-to-energy-plant-at-narela/story-dZuZaGLV3UFQPzU8vmSbyM.html. Accessed March 27, 2018.

29 National Environment Agency. Waste Management. www.nea.gov.sg/energy-waste/ waste-management. Accessed March 27, 2018.

30 State of Delaware. Coastal Zone Act Program. www.dnrece.delaware.gov/admins/cza/ pages/default-aspx. Accessed July 13, 2017

31 Oseinski E, Townsend B. We Must Honor the Coastal Zone Act's Original Purpose. Delaware Voice editorial. May 24, 2017. www.delawareonline.com/story/opinion/ contributors/2017/05/24/must-honor-coastal-zone-acts-original-purpose/101873754/. Accessed March 27, 2018.

32 Horleman C. Commentary: With Coastal Zone Act Update, Delaware Can Strike a Better Balance. *Delaware State News*. May 21, 2017. https://delawarestatenews.net/ opinion/commentary-with-coastal-zone-act-update-delaware-can-strike-a-better-balance/. Accessed March 27, 2018.

33 Min S. Delaware Groups Fighting Proposed Changes to Coastal Zone Act. News-Works. June 6, 2017. https://whyy.org/articles/delaware-groups-fighting-proposed-changes-to-coastal-zone-act/. Accessed March 27, 2018.

34 Min S. Proposed Law Will 'Gut' Delaware's Coastal Zone Act. NewsWorks. May 23, 2017.   https://whyy.org/articles/proposed-law-will-qgutq-delawares-coastal-zone-act/. Accessed March 27, 2018.
35 Gross S. Delaware Coastal Zone Bill Voted Out of House Committee. *The News Journal.* June 7, 2017. www.delawareonline.com/story/news/politics/2017/06/08/ delaware-coastal-zone-bill-voted-out-house-commitee/377768001/. Accessed March 27, 2018.

Index

Page numbers in **bold** denote tables, those in *italics* denote figures.